Introduction to Complex Analysis

Introduction to Complex Analysis

Revised Edition

H. A. PRIESTLEY

CLARENDON PRESS · OXFORD
1990

Oxford University Press, Walton Street, Oxford OX2 6DP

Oxford   New York   Toronto
Delhi   Bombay   Calcutta   Madras   Karachi
Petaling Jaya   Singapore   Hong Kong   Tokyo
Nairobi   Dar es Salaam   Cape Town
Melbourne   Auckland

and associated companies in
Berlin   Ibadan

Oxford is a trade mark of Oxford University Press

Published in the United States
by Oxford University Press, New York

First published 1985
Reprinted 1986, 1988 (with corrections)
Revised edition 1990

British Library Cataloguing in Publication Data
Priestley, H. A. (Hilary A.)
Introduction to complex analysis.—Rev. ed.
1. Calculus. Functions of complex variables
I. Title
515.9

ISBN 0-19-853429-9
ISBN 0-19-853428-0 pbk

Library of Congress Cataloging in Publication Data

Priestley, H. A. (Hilary A.)
Introduction to complex analysis/H. A. Priestley.—Rev. ed. p. cm.
ISBN 0-19-853429-9
ISBN 0-19-853428-0 (pbk.)
1. Functions of complex variables.   I. Title
QA331.7.P75 1990
515'.9—dc20   89-22944 CIP

Printed in Great Britain
by Bookcraft (Bath) Ltd
Midsomer Norton, Avon

Preface to the revised edition

The major change in this revised edition is in the exercises. The main thrust of the text remains unchanged. However some new worked examples have been added at strategic points and a few explanations have been amplified. I am grateful to the students whose occasional blank looks pinpointed places in the text where such insertions were desirable. In addition the bibliography has been expanded.

Many new exercises have been added, mostly in a set of supplementary exercises to be found at the end of the book. The end-of-chapter exercise sets have been revised and now include a greater proportion of elementary exercises, designed to familiarize students with definitions and to test and increase understanding of basic concepts. The Supplementary Exercises are a source of examples for extra practice or revision. In addition they serve to broaden the book's perspective, by introducing some new topics for which I could not, or did not wish to, find space in the text. I claim little originality for the 'new' exercises. Many are very well worn indeed and others have been adapted from past Oxford University examination papers.

My thanks go to all who spotted misprints in the original and subsequent printings, both those I cajoled and bribed to hunt mistakes and those who, unsolicited, took the trouble to write to me. In particular I must thank Dr Irene Ault, who kindly undertook to check the new and amended exercises.

Oxford
November 1989   H. A. P.

Preface to the first edition

This is a textbook for a first course in complex analysis. It aims to be practical without being purely utilitarian and to be rigorous without being over-sophisticated or fussy. The power and significance of Cauchy's theorem—the centre-piece of complex analysis—is, I believe, best revealed initially through its applications. Consequently, emphasis has been put on those parts of the subject emanating from Cauchy's integral formula and Cauchy's residue theorem. This does not mean that the geometrical and topological aspects of complex analysis have been neglected, merely that it is recognized that a full appreciation of such concepts as index only comes with experience. Thus the chapters in which these important foundations are discussed are written in such a way that the student may at a first reading easily extract what he needs to proceed to the applications. He is, naturally, encouraged to return later in search of a deeper understanding.

The book is a metamorphosis of a set of notes in the series produced by the Mathematical Institute of the University of Oxford. As student opinion dictated it should, it betrays its previous incarnation—notably in its brevity and its style. Essential ideas are not submerged in a welter of details, material is locally arranged for ease of reference, and by-ways (however fascinating) are left unexplored. Advanced and specialized topics have been ruthlessly excluded. So, for example, analytic continuation and special functions receive only passing mention; a satisfactory treatment of either would have made unacceptable demands on the reader. Applied mathematicians have been provided with a thorough account of applicable complex analysis, but specific physical problems are not discussed. A chapter on Fourier and Laplace transforms has been included. This is used to show off the techniques of residue calculus developed in the preceding chapters. It is also designed as a self-contained introduction to transform methods (and so strays somewhat beyond the confines of complex analysis), but does not purport to be a comprehensive survey of transform theory.

Some prior acquaintance with complex numbers is assumed. Apart from this, the only prerequisite is a course in elementary real analysis involving some exposure to ε–δ proofs. Many analysis and calculus texts cover the required background. I have taken K. G. Binmore, *Mathematical analysis: a straightforward introduction* [4] as my basic reference since it has the merit of containing the appropriate minimum of topics and of having the same philosophy as the present book. Those concepts in complex analysis which transfer, *mutatis mutandis*, from real analysis (continuity, etc.) are treated very briefly. Few students welcome, or benefit from, a detailed presentation of essentially familiar technical material. Also, the more time spent in these foothills, the less time is available for exploring the novel and spectacular terrain surrounding Cauchy's theorem.

Not all readers will have the same mathematical background. To allow for this I have adopted the convention that text enclosed in square brackets should be heeded by anyone to whom it makes sense but may safely be ignored by others. These occasional bracketed comments concern, for example, certain results in topology. It is accepted practice for texts on complex analysis to work with the Riemann integral rather than the Lebesgue integral. This is irritating for those who have graduated to the latter and confusing for those (Oxford students in particular) who are never taught the former. A dual approach is adopted here. To understand the book the reader needs a rudimentary knowledge of either Riemann integration or Lebesgue integration; signposts are provided for the followers of each theory.

Certain theorems have been designated with the customary proper names, but I have otherwise made no attempt to attribute theorems or proofs. Also, the subject has been so well worked over that I do not claim any originality for methods, examples, or exercises I happen never to have seen elsewhere. Among the books I have found particularly influential have been those by W. Rudin [7] and A. F. Beardon [3].

My preliminary notes on complex analysis evolved over about ten years. The first version of these was based on some notes produced by Dr Ida Busbridge. She had earlier introduced me to 'complex variable', and I gratefully acknowledge my debt to her. It is also a pleasure to thank those colleagues in Oxford and elsewhere who directly or indirectly have had an influence on the book. However, my special thanks go to Dr Christine Farmer of London University; she has been involved with this project since its inception and has read draft after draft with care and patience. Her constructive criticisms have been invaluable and her pencilled question marks unerring. Finally, thanks are due to Professor Michael Adams and Professor Michael Albert for their help with proof-reading, and to the staff of the Oxford University Press for encouraging me to write the book and for their assistance during its production.

Oxford
March 1985   H. A. P.

Contents

Notation and terminology

We use $\mathbb{Z}$, $\mathbb{R}$, and $\mathbb{C}$ to denote the set of integers, the set of real numbers, and the set of complex numbers, respectively. Standard terms and symbols relating to sets and mappings have their conventional meanings. The following notation, which may not be universally familiar, is also used. Given sets A and B, the set $\{a \in A : a \notin B\}$ is denoted by $A \backslash B$, and given a mapping $f : A \rightarrow B$, we write the image set $\{f(a) : a \in A\}$ as $f(A)$. The symbol $:=$ denotes 'equals by definition'; it is used to stress that an equation is defining something and also as a convenient shorthand. We signify the end of a proof or solution by $\square$. Finally, we adopt the Bourbaki dangerous bend symbol $\gtrless$ to warn of a common pitfall.

When we extend such concepts as differentiability from a real to a complex setting we shall sometimes transfer secondary vocabulary and notation without comment. For example, once $f'(z)$ has been defined, we credit the reader with enough common sense to deduce what is meant by $f''(z)$ and $f^{(n)}(z)$.

As indicated in the preface, any comment in the text enclosed in square brackets is aimed just at those readers who happen to have the knowledge necessary to understand it.

1 The complex plane

Complex analysis has its roots in the algebraic, geometric, and topological structure of the complex plane. This chapter explores these foundations. It is assumed that the reader has previously been introduced to complex numbers, and has had some practice in manipulating them. Consequently, the first part of the chapter is designed to be a refresher course. It contains a summary of basic properties, presented without undue formality.

Complex numbers

1.1 Complex numbers and their representation

A complex number z is specified by a pair of real numbers x and y; we write $z = x + iy$, where i (sometimes known as j) is a fixed symbol. (The arithmetical rules given below force $i^2 = -1$.) The set of complex numbers is denoted by $\mathbb{C}$. Two elements $x + iy$ and $u + iv$ of $\mathbb{C}$ are, by definition, equal if and only if $x = u$ and $y = v$. This allows us, given $z = x + iy \in \mathbb{C}$, unambiguously to define x to be the real part of z and y to be the imaginary part. We use the customary abbreviations: x for $x + i0$, iy (or yi) for $0 + iy$, and i for $0 + i1$. The first of these implies that we regard the set of real numbers, $\mathbb{R}$, as a subset of $\mathbb{C}$. The terminology here is standard, but a legacy from the past: complex numbers are not complex, nor imaginary numbers imaginary.

It is convenient to represent complex numbers geometrically as points of a plane (the complex plane). We equip the plane $\mathbb{R}^2$ in the usual way with cartesian coordinate axes and identify $z = x + iy$ with $(x, y) \in \mathbb{R}^2$. Alternatively we may use polar coordinates and write, for $(x, y) \in \mathbb{R}^2$, $x = r \cos \theta$ and $y = r \sin \theta$.

For a given complex number $z \in \mathbb{C}$, we thus have the following.

Cartesian representation: $z = x + iy$; we write $\operatorname{Re} z = x$ and $\operatorname{Im} z = y$. The *modulus* of z is defined to be

$$|z| = \sqrt{(x^2 + y^2)}$$

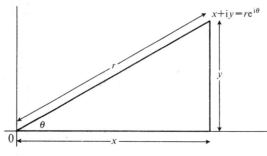

Fig. 1.1

(where the positive square root is taken). Notice that

$$z = 0 \Leftrightarrow \operatorname{Re} z = \operatorname{Im} z = 0 \Leftrightarrow |z| = 0.$$

Polar representation: $z = r\cos\theta + ir\sin\theta = re^{i\theta}$ (where, pro tem., we simply regard $e^{i\theta}$ as shorthand for $\cos\theta + i\sin\theta$). We have $|z| = r$. For $z = 0$, the angle θ can be chosen arbitrarily. For $z \neq 0$, it is not uniquely determined: because of the periodicity of the functions cosine and sine, θ is only determined up to an integer multiple of 2π. We call any value of θ with $z = |z|e^{i\theta}$ *an argument* of z. This non-uniqueness, which may appear here merely to be an inconvenience, turns out to have far-reaching consequences (see 2.18 and 4.15–4.19).

1.2   The algebraic structure of the complex numbers

By extension of the corresponding operations for real numbers, addition and multiplication are defined in $\mathbb{C}$ by

$$(x + iy) + (u + iv) = (x + u) + i(y + v)$$

and

$$(x + iy)(u + iv) = (xu - yv) + i(xv + yu).$$

Taking $x = u = 0$ and $y = v = 1$, we obtain the identity $i^2 = -1$.

Routine checking shows that the same arithmetical rules apply in $\mathbb{C}$ as in $\mathbb{R}$. For z_1, z_2, and $z_3 \in \mathbb{C}$, we have

commutative laws

$$z_1 + z_2 = z_2 + z_1 \quad \text{and} \quad z_1 z_2 = z_2 z_1,$$

associative laws

$$z_1 + (z_2 + z_3) = (z_1 + z_2) + z_3 \quad \text{and} \quad z_1(z_2 z_3) = (z_1 z_2)z_3,$$

and the distributive law

$$z_1(z_2 + z_3) = z_1 z_2 + z_1 z_3.$$

For all $z \in \mathbb{C}$, $0 + z = z$, $1z = z$. Further, given $z = x + iy$, there exists $-z := (-x) + i(-y)$ such that $z + (-z) = 0$ and, so long as $z \neq 0$, there exists $1/z := x(x^2 + y^2)^{-1} + i[-y(x^2 + y^2)^{-1}]$ such that $z(1/z) = 1$. In a mathematical nutshell, $\mathbb{C}$ forms a field. As usual, we write $z - w$ for $z + (-w)$ and z/w for $z(1/w)$ $(w \neq 0)$.

It is worth noting that, while addition is most conveniently expressed using the cartesian representation, the neatest formula for multiplication is in terms of the polar representation. The product of $z = re^{i\theta}$ and $w = Re^{i\phi}$ is given by $zw = rRe^{i(\theta + \phi)}$. Hence comes de Moivre's theorem:

$$(\cos \theta + i \sin \theta)^n = \cos n\theta + i \sin n\theta \quad (\theta \in \mathbb{R}).$$

1.3 Complex conjugation

Given $z = x + iy$, the *complex conjugate* of z is defined to be $\bar{z} := x - iy$. In polar form, $z = re^{i\theta}$ implies $\bar{z} = re^{-i\theta}$. It is easily verified that the following identities hold for all z and $w \in \mathbb{C}$:

(1) $\bar{\bar{z}} = z$;
(2) $2 \operatorname{Re} z = z + \bar{z}$, $\quad 2i \operatorname{Im} z = z - \bar{z}$;
(3) $|\bar{z}| = |z|$, $\quad z\bar{z} = |z|^2$;
(4) $\overline{z + w} = \bar{z} + \bar{w}$, $\quad \overline{zw} = \bar{z}\bar{w}$.

From (3) we see that $|zw| = |z| \, |w|$. In general $|z + w| \neq |z| + |w|$. However, important inequalities link modulus and addition.

1.4 Inequalities

For all z and $w \in \mathbb{C}$:

(1) $|\operatorname{Re} z| \leqslant |z|$, $\quad |\operatorname{Im} z| \leqslant |z|$;
(2) $|z + w| \leqslant |z| + |w|$ (the triangle inequality);
(3) $|z + w| \geqslant ||z| - |w||$.

Proof. (1) is immediate from the relations

$$|z|^2 = (\operatorname{Re} z)^2 + (\operatorname{Im} z)^2, \quad |z| \geqslant 0.$$

To prove (2), observe that

$$
\begin{aligned}
|z + w|^2 &= (z + w)(\bar{z} + \bar{w}) && \text{(by 1.3 (3) and (4))} \\
&= |z|^2 + |w|^2 + (w\bar{z} + z\bar{w}) && \text{(by 1.3 (3))} \\
&= |z|^2 + |w|^2 + 2 \operatorname{Re}(z\bar{w}) && \text{(by 1.3 (1) and (2))} \\
&\leqslant |z|^2 + |w|^2 + 2 |z| \, |w| && \text{(by (1) and 1.3 (3) and (4))} \\
&= (|z| + |w|)^2.
\end{aligned}
$$

Since $|z + w| \geqslant 0$ and $|z| + |w| \geqslant 0$ we deduce (2).

For real numbers a and b, the inequality $|a| \leqslant b$ holds if and only if $a \leqslant b$ and $-a \leqslant b$. Hence (3) is satisfied provided the two inequalities $|z+w| \geqslant |z| - |w|$ and $|z+w| \geqslant |w| - |z|$ hold. But by (2) we have

$$|z| = |z + w - w| \leqslant |z + w| + |w| \quad \text{and} \quad |w| = |z + w - z|$$
$$\leqslant |z + w| + |z|,$$

whence (3) follows. ☐

Note All of the above inequalities concern complex numbers but are between real numbers ($|z|$, Re z, etc.). It is important to realize that no meaning has been assigned to an inequality between complex numbers. Indeed, it is not possible to define an ordering on $\mathbb{C}$ in which any two elements are comparable, such as exists on $\mathbb{R}$ (see Exercise 1.7). Whenever inequalities appear henceforth, the quantities involved are to be assumed to be real. Thus $w \geqslant 0$ means that w is a real number which is also non-negative. In complex analysis, abuse of inequalities is perhaps the most common type of error perpetrated by beginners. You have been warned!

1.5 Subsets of the complex plane

The modulus of $z = x + iy$ has a geometric interpretation. It gives the distance of the point (x, y) from the origin of coordinates. More generally, $|z - w|$ is the distance between points z and w in the plane, and the triangle inequality $|z + w| \leqslant |z| + |w|$ can be interpreted as the assertion that the length of one side of any triangle does not exceed the sums of the lengths of the other two sides.

In the opposite direction, we can give a variety of descriptions of geometric entities in terms of complex numbers.

(1) Lines (i) the real axis is given by any of the equations
  (a) Im $z = 0$;
  (b) $z = \bar{z}$;
  (c) $|z - i| = |z + i|$, or more generally $|z - a| = |z - \bar{a}|$ (where Im $a \neq 0$).
The first of these may be taken as the definition. The second is clearly equivalent to it. The conjugate $\bar{a}$ of a is represented in the plane by the reflection of a in the real axis and (c) is simply saying that the real axis consists of points equidistant from a and $\bar{a}$.

(ii) The imaginary axis can be similarly described. In particular it is given by either of the equations Re $z = 0$ or $|z - 1| = |z + 1|$.

(iii) The perpendicular bisector of the line segment joining the points a and b is given by $|z - a| = |z - b|$. The equation of any line can, for suitable a and b, be expressed in this convenient form.

(2) Line segments For a and $b \in \mathbb{C}$, the line segment with end-points a and b is given by

$$[a, b] = \{(1 - t)a + tb : 0 \leqslant t \leqslant 1\}.$$

(3) Circles The circle centre $a \in \mathbb{C}$ and radius $r > 0$ has equation $|z - a| = r$. A typical point on the circle has the form $z = a + re^{i\theta}$ ($\theta \in \mathbb{R}$). The circle centre 0 and radius 1 is called the unit circle. Another way of specifying circles is discussed in Chapter 10.

Descriptions of loci such as lines and circles lead easily to ways of describing various plane sets, as illustrated below.

1.6 Examples

(1) The upper half-plane (excluding the real axis) is $\{z : \operatorname{Im} z > 0\}$ or equivalently $\{z : |z - i| < |z + i|\}$ (viz. the set of points closer to i than to $-$i).

(2) $\{z : 1 \leqslant |z| \leqslant 2\}$ describes the annulus shown in Fig. 1.2(i); here the boundary circles are included.

(3) The set $S = \{z : |z + \frac{2}{3}| < \frac{1}{3}, \text{ and } |z + \frac{1}{3}| < |z + 1|\}$ consists of those points which lie both inside the circle centre $-\frac{2}{3}$, radius $\frac{1}{3}$, and closer to $-\frac{1}{3}$ than to -1. Hence S is the shaded semicircle shown in Fig. 1.2(ii); in this case the boundary is not included.

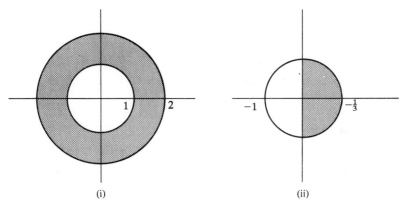

(i)   (ii)

Fig. 1.2

Open and closed sets in the complex plane

Complex analysis has a vital geometric facet, from which it derives much of its character. The geometry of the plane is inextricably bound up with its topological structure, described by the open sets we define below. We assume no prior acquaintance with topology, and include in this section only the bare minimum for our immediate needs. Those who have studied the rudiments of topology in Euclidean space should find nothing unfamiliar here.

1.7 Discs in the complex plane

In elementary real analysis the subsets of $\mathbb{R}$ with which one is most concerned are intervals. Bounded open intervals, that is, non-empty sets $(c, d) := \{x \in \mathbb{R} : c < x < d\}$ $(c, d \in \mathbb{R})$, underlie the definitions of limits and continuity, while bounded closed intervals (of the form $[c, d] := \{x \in \mathbb{R} : c \leqslant x \leqslant d\}$ $(c, d \in \mathbb{R})$) are the sets on which continuous functions have especially good behaviour (boundedness, intermediate-value property, etc.). Any bounded open (closed) interval in $\mathbb{R}$ can be expressed in the form $\{x \in \mathbb{R} : |x - a| < r\}$ $(\{x \in \mathbb{R} : |x - a| \leqslant r\})$, for some a and $r \in \mathbb{R}$. In $\mathbb{C}$, we define, analogously, the *open disc* centre a, radius $r > 0$, to be

$$D(a; r) := \{z \in \mathbb{C} : |z - a| < r\},$$

the *closed disc* centre a, radius $r > 0$, to be

$$\bar{D}(a; r) := \{z \in \mathbb{C} : |z - a| \leqslant r\},$$

and the *punctured disc* centre a, radius $r > 0$, to be

$$D'(a; r) := \{z \in \mathbb{C} : 0 < |z - a| < r\}.$$

The adjectives 'open' and 'closed' are here applied to very special subsets of $\mathbb{C}$. Definitions 1.8 and 1.10 extend the use of these terms, in a consistent way (see Examples 1.9(1), 1.12(1)).

1.8 Definition

A set $S \subseteq \mathbb{C}$ is *open* if, given $z \in S$, there exists $r > 0$ (depending on z) such that $D(z; r) \subseteq S$.

Informally, S is open if, from any given point in S, there is room to move some fixed positive distance in any direction without straying outside S; how large this distance can be will vary from

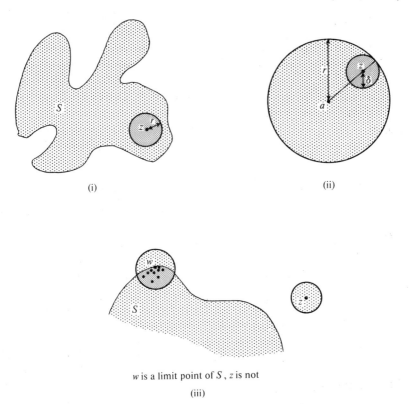

(i)

(ii)

w is a limit point of S, z is not

(iii)

Fig. 1.3

one point to another (see Fig. 1.3(i)). It is a need for such 'elbow room' that dictates that the sets in so many of our later theorems be open. [The open sets above determine the usual metric topology on $\mathbb{C}$ defined by the modulus function.]

1.9 Examples

(1) For any $a \in \mathbb{C}$ and $r > 0$, $D(a; r)$ is an open set. To see this, fix $z \in D(a; r)$ and note that $D(z; \delta) \subseteq D(a; r)$ provided $0 < \delta \leq r - |z - a|$ (see Fig. 1.3(ii)).

(2) The quadrant $Q = \{z \in \mathbb{C} : \operatorname{Re} z > 0, \operatorname{Im} z > 0\}$ is open since, for $z \in Q$, $D(z; \delta) \subseteq Q$ whenever $0 < \delta < \min\{\operatorname{Re} z, \operatorname{Im} z\}$. Similarly, rectangles, sectors, and annuli, when specified by strict inequalities, are open.

1.10 Definitions

(1) A set $S \subseteq \mathbb{C}$ is *closed* if $\mathbb{C} \backslash S := \{z \in \mathbb{C} : z \notin S\}$ is open.
(2) A point $z \in \mathbb{C}$ is a *limit point* of $S \subseteq \mathbb{C}$ if $D'(z; r) \cap S \neq \emptyset$ for every $r > 0$.
(3) The *closure* $\bar{S}$ of $S \subseteq \mathbb{C}$ is the union of S and its limit points.

In (2) the disc is a punctured one to preclude points of S being automatic limit points. For z to qualify as a limit point of S it must have other members of S clustering round it; see Fig. 1.3(iii).

1.11 Proposition

A set $S \subseteq \mathbb{C}$ is closed if and only if S contains all its limit points.

Proof. Note that for $z \notin S$, $D(z; r) \cap S = D'(z; r) \cap S$. From the definitions,

S is closed $\Leftrightarrow \mathbb{C} \backslash S$ is open

$\qquad \Leftrightarrow$ given $z \notin S$, there exists $r > 0$ such that $D(z; r) \subseteq \mathbb{C} \backslash S$

$\qquad \Leftrightarrow$ given $z \notin S$, there exists $r > 0$ such that $D'(z; r) \cap S = \emptyset$

$\qquad \Leftrightarrow$ no point of $\mathbb{C} \backslash S$ is a limit point of S. $\qquad \square$

1.12 Examples

(1) Any closed disc $\bar{D}(a; r)$ is a closed set. To prove this, fix $z \notin \bar{D}(a; r)$. Then $D(z; \delta) \subseteq \mathbb{C} \backslash D(a; r)$ whenever $0 < \delta \leq |z - a| - r$. The set of limit points of $D(a; r)$ is $\bar{D}(a; r)$, so the closure of $D(a; r)$ is $\bar{D}(a; r)$.
(2) Half-planes, quadrants, annuli, etc., when specified by weak inequalities ($\leq$), are closed.
(3) The set $S = \{z : 1 \leq |z| < 2\}$ is neither open nor closed, since no disc $D(1; r) \subseteq S$ and no disc $D(2; r) \subseteq \mathbb{C} \backslash S$. Sets are not like doors: non-open sets do not have to be closed. The concepts open and closed are related by complementation, not negation.
(4) Let $S = \{(-1)^n(1 + 1/n) : n = 1, 2, \ldots\}$. Then S has ± 1 as its limit points. It is neither open nor closed.

1.13 Definitions

A subset S of $\mathbb{C}$ is said to be *bounded* if there exists a constant M such that $|z| \leq M$ for all $z \in S$. A set which is both closed and bounded will be called *compact*.

Note that, in particular, line segments $[a, b]$ in $\mathbb{C}$ and closed discs $\bar{D}(a;r)$ are compact. [The sets we have called compact are exactly those which satisfy the usual open covering definition of compactness, thanks to the Heine–Borel theorem. Save in Theorem 3.23, where we proceed *ad hoc*, we shall not need to work with open coverings.]

Limits and continuity

This section contains the technical foundations of analysis in the complex plane. It deals only with those concepts and results which transfer in an entirely straightforward manner from elementary real analysis. We assume the reader is already familiar with limits and continuity in the real case, as presented, for example, in Binmore [4]. The definitions which follow mimic those for real sequences and functions, with open discs in $\mathbb{C}$ replacing open intervals in $\mathbb{R}$.

1.14 Definitions

(1) A (*complex*) *sequence* $\langle z_n \rangle$ is an assignment of a complex number z_n to each natural number $n = 1, 2, \ldots$. We occasionally need to allow a different set of values of n; notation such as $\langle z_n \rangle_{n \geqslant 0}$ should be self-explanatory. The sequence $\langle z_n \rangle$ is *bounded* if there is a finite constant M such that $|z_n| \leqslant M$ for all n.

(2) The sequence $\langle z_n \rangle$ *converges* to the limit a (in symbols, $z_n \to a$) as $n \to \infty$ if, given $\varepsilon > 0$, there exists a natural number N (depending on ε) such that

$$n \geqslant N \quad \text{implies} \quad |z_n - a| < \varepsilon.$$

(3) The sequence $\langle w_n \rangle$ is a *subsequence* of the sequence $\langle z_n \rangle$ if there exist natural numbers $n_1 < n_2 < \ldots$ such that $w_k = z_{n_k}$ ($k = 1, 2, \ldots$).

(4) Let $f : S \to \mathbb{C}$ be a function defined on a set $S \subseteq \mathbb{C}$ and let $a \in \bar{S}$. Then $\lim_{z \to a} f(z) = w$ (in alternative notation, $f(z) \to w$ as $z \to a$) if, given $\varepsilon > 0$, there exists $\delta > 0$ (depending on a and ε) such that

$$0 < |z - a| < \delta \quad \text{implies} \quad |f(z) - w| < \varepsilon.$$

(5) Let $f : S \to \mathbb{C}$ be a function. Then f is *continuous* at $a \in S$ if given $\varepsilon > 0$, there exists $\delta > 0$ (depending on a and ε) such that

$$z \in D(a; \delta) \cap S \quad \text{implies} \quad |f(z) - f(a)| < \varepsilon;$$

f is *continuous on* S if it is continuous at each $a \in S$. (Continuous functions can be more elegantly characterized in terms of open sets, but we require only the ε–δ definition.)

The algebra of complex limits (sums, products, etc.), and other elementary properties, can be developed, both for sequences and for functions, exactly as in the real case. We shall assume this done, and feel free to use the results. One caveat is however necessary. Proofs which depend on the order structure of $\mathbb{R}$ do not transfer directly to $\mathbb{C}$. It is therefore useful to have available the following lemma linking convergence in $\mathbb{C}$ with convergence in $\mathbb{R}$. Both parts are easily derived from the definitions, using 1.4(1).

1.15 Lemma

(1) Let $\langle z_n \rangle$ be a complex sequence. Then $\langle z_n \rangle$ converges if and only if both the real sequences $\langle \text{Re } z_n \rangle$ and $\langle \text{Im } z_n \rangle$ converge, and if $z_n \to a$, then $\text{Re } z_n \to \text{Re } a$ and $\text{Im } z_n \to \text{Im } a$.

(2) Let $f : S \to \mathbb{C}$ be a function and write $f = \text{Re } f + i \text{ Im } f$, where $\text{Re } f$, $\text{Im } f$ are real-valued functions, given by

$$(\text{Re } f)(z) = \text{Re } f(z), \quad (\text{Im } f)(z) = \text{Im } f(z) \quad (z \in S).$$

Then for any $a \in \bar{S}$, $\lim_{z \to a} f(z)$ exists if and only if both $\lim_{z \to a} \text{Re } f(z)$ and $\lim_{z \to a} \text{Im } f(z)$ exist. In this case, if $f(z) \to w$, then $(\text{Re } f)(z) \to \text{Re } w$ and $(\text{Im } f)(z) \to \text{Im } w$.

We use the lemma to derive the next theorem from its real counterpart (given, for example, in Binmore [4], 5.10). This theorem will be used to obtain two further results which are needed in the course of proving some important theorems later on.

1.16 Theorem

Any bounded sequence in $\mathbb{C}$ has a convergent subsequence.

Proof. Let $\langle z_n \rangle$ be a sequence with $|z_n| \leqslant M$ for all n. Then (by 1.4(1)) $|\text{Re } z_n| \leqslant M$, so $\langle \text{Re } z_n \rangle$ is a bounded sequence in $\mathbb{R}$. Hence there exist $n_1 < n_2 < \ldots$ such that $\langle \text{Re } z_{n_k} \rangle_{k \geqslant 1}$ converges. The sequence $\langle \text{Im } z_{n_k} \rangle_{k \geqslant 1}$ is a bounded real sequence and we can choose natural numbers $m_j = n_{k_j}$ with $m_1 < m_2 < \ldots$ so that $\langle \text{Im } z_{m_j} \rangle_{j \geqslant 1}$ converges. As a subsequence of $\langle \text{Re } z_{n_k} \rangle_{k \geqslant 1}$, the sequence $\langle \text{Re } z_{m_j} \rangle_{j \geqslant 1}$ must converge too. Lemma 1.15 shows that $\langle z_{m_j} \rangle_{j \geqslant 1}$ provides a convergent subsequence of $\langle z_n \rangle$. ☐

1.17 Corollary (The Bolzano–Weierstrass theorem)

Any infinite compact subset S of $\mathbb{C}$ has a limit point in S.

Proof. By Definition 1.13, S is bounded and closed. Select a sequence $\langle z_n \rangle$ with the points z_n distinct and belonging to S. Theorem 1.16 asserts that $\langle z_n \rangle$ has a subsequence which converges. If z is the limit of such a subsequence, z is a limit point of S. Because S is closed, it contains z (by Proposition 1.11). □

1.18 Theorem

Let S be a compact subset of $\mathbb{C}$ and $f: S \to \mathbb{C}$ a continuous function. Then:
(1) f is bounded, that is, there exists a finite constant M such that $|f(z)| \leq M$ for all $z \in S$;
(2) f attains its bounds, that is, there exist z_1 and $z_2 \in S$ such that

$$|f(z_1)| \leq |f(z)| \leq |f(z_2)| \quad \text{for all} \quad z \in S.$$

Proof. With 1.17 available, we can proceed as is customary in the real case; see for example [4], 9.11, 9.12. □

We conclude by recording, again for future use, a theorem from real analysis which is an immediate corollary of the Intermediate-value theorem ([4], 9.10).

1.19 Theorem

Let f be an integer-valued continuous function on a compact subinterval of $\mathbb{R}$. Then f is constant.

Exercises

1. Express each of the following in polar form: i, $1-i$, $\sqrt{3}-i$, $(1-i)/(1+\sqrt{3}i)$, $(\cos\frac{1}{2} - i\sin\frac{1}{2})^2$.

2. Express the following in the form $x + iy$ $(x, y \in \mathbb{R})$: $e^{4\pi i/3}$, $e^{5\pi i/6}$, $(1+i)^{-3}$, $(1+2i)/(2+i)$, $[\alpha^2(\alpha-1)^2]^{-1}$, where $\alpha^3 = -1$, $\alpha \neq -1$.

3. Evaluate, for $n = 1, 2, 3, \ldots$,

(i) i^n,   (ii) $\left(\dfrac{1-i}{1+i}\right)^n$,   (iii) $(1+i)^n + (1-i)^n$.

4. Prove that, for $z \in \mathbb{C}$, $|z| \leq |\text{Re } z| + |\text{Im } z| \leq \sqrt{2}|z|$. Give examples to show that either inequality may be an equality.

5. Prove that, for z and w in $\mathbb{C}$,

$$|1 - \bar{z}w|^2 - |z - w|^2 = (1 - |z|^2)(1 - |w|^2).$$

Deduce that, if $|z| < 1$ and $|w| < 1$,

$$\left|\frac{z - w}{1 - \bar{z}w}\right| < 1.$$

6. Let $z = re^{i\theta}$ and $w = Re^{i\phi}$, where $0 \leqslant r < R$. Prove that

$$\text{Re}\left(\frac{w+z}{w-z}\right) = \frac{|w|^2 - |z|^2}{|w-z|^2} = \frac{R^2 - r^2}{R^2 - 2Rr\cos(\theta - \phi) + r^2}.$$

(These formulae for the *Poisson kernel* are needed in Chapter 10.)

7. The usual order relation $>$ on $\mathbb{R}$ satisfies
(i) $x \neq 0 \Rightarrow x > 0$ or $-x > 0$, but not both, and
(ii) $x, y > 0 \Rightarrow x + y > 0$ and $xy > 0$.
Show that there does not exist a relation $>$ on $\mathbb{C}$ satisfying (i) and (ii). (Hint: consider i.)

8. Describe the following sets geometrically.
(i) $\{z : 1 < \text{Im}(z + i) < 2\}$, (iii) $\{z : |z - i| < |z - 1|\}$,
(ii) $\{z : |z + 2i| \geqslant 2\}$, (iv) $\{z = |z|e^{i\theta} : -\pi < \theta < \pi/2\}$.

9. Prove that any punctured disc $D'(a; r)$ is an open set. What are its limit points? What is its closure?

10. Let G be an open set in $\mathbb{C}$. Which of the following sets are open:
(i) $\{z : \bar{z} \in G\}$, (ii) $\{\text{Re } z : z \in G\}$, (iii) $\{z : z \in G \text{ or } \bar{z} \in G\}$,
(iv) $\{z \in G : \text{Im } z > 0\}$?

11. Describe geometrically each of the following sets. Which are open, which are closed, and which are compact? Find the closures of the non-closed sets.
(i) $\{z : |z - 1 - i| = 1\}$, (vi) $\{z : |z - 1| < 1, |z| = |z - 2|\}$,
(ii) $\{z : |z - 1 + i| \geqslant |z - 1 - i|\}$, (vii) $\{z : |z - 2| > 3, |z| < 2\}$,
(iii) $\{z : |z + i| \neq |z - i|\}$, (viii) $\{z : |z^2 - 1| < 1\}$,
(iv) $\{z = |z|e^{i\theta} : \frac{1}{4}\pi < \theta \leqslant \frac{3}{4}\pi\}$, (ix) $\{z : |z|^2 > z + \bar{z}\}$,
(v) $\{z : \text{Re } z < 1 \text{ or } \text{Im}(z - 1) \neq 0\}$, (x) $\{z : \text{Im}[(z + i)/2i] < 0\}$.

12. Which of the following complex sequences converge:
(i) $\left\langle \frac{1}{n}i^n \right\rangle$, (ii) $\left\langle \frac{(-1)^n n}{n + i} \right\rangle$, (iii) $\left\langle \frac{n^2 + in}{n^2 + i} \right\rangle$?

13. Suppose $\langle z_n \rangle$ is a complex sequence of distinct points converging to z. Prove that z is the unique limit point of the set $\{z_n : n = 1, 2, \ldots\}$.

14. For each of the following choices of $f(z)$, either obtain $\lim_{z \to 0} f(z)$ or prove that the limit fails to exist:
(i) $\frac{|z|^2}{z}$, (ii) $\frac{\bar{z}}{z}$, (iii) $\frac{z + 1}{|z| - 1}$, (iv) $\frac{(\text{Re } z)(\text{Im } z)}{|z|}$.

15. Prove that f is continuous on $\mathbb{C}$ when (i) $f(z) = \bar{z}$, (ii) $f(z) = \text{Im } z$, (iii) $f(z) = \text{Re } z^2$.

16. Define a function f by $f(z) = z/(1 + |z|)$. Show that f is continuous and that it maps $\mathbb{C}$ one-to-one onto $D(0; 1)$.

2 Holomorphic functions and power series

Chapter 1 did not go far enough to reveal the true flavour of complex analysis. Continuous functions play only an ancillary and technical role in the subject. Much more important are the holomorphic functions which this chapter introduces. Loosely, holomorphic means differentiable. The formal definition of holomorphy, given in 2.2, is restrictive enough to lead to powerful and elegant theorems, yet wide enough to allow a wealth of practical applications. Power series are central to the development of the theory. As Theorems 2.12 and 5.9 will show, they define, and can be used to represent, holomorphic functions.

That part of Chapter 10 dealing with examples of holomorphic mappings can profitably be studied immediately after this chapter.

Holomorphic functions

2.1 Functions

In Chapter 1 we took for granted the concept of a function. Formally, given $S \subseteq \mathbb{C}$, a mapping $f : S \to \mathbb{C}$ which assigns to each $z \in S$ a unique complex number $f(z)$ is called a *complex-valued function*, or simply a *function*, on S. Reference back to this fundamental definition is not merely belated pedantry. This is simply an opportune point at which to emphasize the inherent 'one-valuedness' of a function. (Later we deal also with what we call multifunctions: a *multifunction* is a rule assigning a subset of $\mathbb{C}$ (finite or infinite) to each element of its domain set S.)

Strictly, given a function f, we should distinguish between f (the mapping) and $f(z)$ (the image of the point z under f). However, where it would be cumbersome to do otherwise, we allow $f(z)$ to denote the function and write, for example 'z^2' in place of 'the function f defined by $f(z) = z^2$'. We also adopt the notation $z \mapsto w = f(z)$ to indicate that z is mapped by f to w.

2.2 Definitions

(1) A function f defined on an open subset G of $\mathbb{C}$ is *differentiable* at $z \in G$ if

$$\lim_{h \to 0} \frac{f(z+h) - f(z)}{h}$$

exists. When the limit exists it is denoted by $f'(z)$ or df/dz.

(2) A function which is differentiable at every point of an open set G is *holomorphic* in G. The set of functions holomorphic in G is denoted by $\mathrm{H}(G)$.

(3) If S is any subset of $\mathbb{C}$, f is holomorphic in S if $f \in \mathrm{H}(G)$ for some open set G containing S.

2.3 Remarks

(1) Suppose f is differentiable at a point z of an open set G in which f is defined. Then, for $z + h \in G$,

$$f(z+h) = f(z) + hf'(z) + h\varepsilon(h) \quad \text{where} \quad \varepsilon(h) \to 0 \quad \text{as} \quad h \to 0.$$

This is immediate on writing, for $h \neq 0$, $\varepsilon(h)$ for

$$\frac{f(z+h) - f(z)}{h} - f'(z),$$

and shows that $f(z+h) \to f(z)$ as $h \to 0$. Hence differentiability of f at z implies continuity of f at z. We have established the technically useful fact that, if f is holomorphic in a set S, then f is continuous on S.

(2) Note the role played by open sets in 2.2. Provided G is open, whenever $z \in G$ there exists $r > 0$ such that $z + h \in G$ for all h with $|h| < r$. This has the effect that in the computation of the limit defining $f'(z)$, $z + h$ is free to approach z from any direction as $h \to 0$. For f to be differentiable at z, the value of the limit must be independent of the manner in which $h \to 0$. We exploit this in the derivation of the following result by equating the expressions obtained for $f'(z)$ on taking (i) h real and (ii) h purely imaginary.

2.4 Theorem (the Cauchy–Riemann equations)

Let f be defined on an open set G and be differentiable at $z = x + iy \in G$. Let

$$f(z) = u(x, y) + iv(x, y),$$

where u and v are real-valued functions on G (regarded as a subset of $\mathbb{R}^2$). Then u and v have first order partial derivatives at (x, y) (denoted u_x, u_y, v_x, v_y) and these satisfy the *Cauchy–Riemann equations*

$$u_x = v_y, \qquad u_y = -v_x.$$

Proof. From Definition 2.2(1),

$$f'(z) = \lim_{h \to 0} \frac{f(z+h) - f(z)}{h}.$$

Hence

$$f'(z) = \lim_{\substack{h \to 0 \\ h \in \mathbb{R}}} \left(\frac{u(x+h, y) - u(x, y)}{h} + i\, \frac{v(x+h, y) - v(x, y)}{h} \right) = u_x + i v_x$$

and

$$f'(z) = \lim_{\substack{h \to 0 \\ h = ik \\ k \in \mathbb{R}}} \left(\frac{u(x, y+k) - u(x, y)}{ik} + \frac{v(x, y+k) - v(x, y)}{k} \right) = \frac{1}{i} u_y + v_y.$$

Equating the two expressions for $f'(z)$ gives

$$u_x + i v_x = -i u_y + v_y.$$

Equating real and imaginary parts we obtain

$$u_x = v_y, \qquad u_y = -v_x. \qquad \square$$

2.5 Example

Let $f(z) = |z|$ on $G = \mathbb{C}$. Prove f is not differentiable at any point.

Solution. In the notation of Theorem 2.4,

$$u(x, y) = (x^2 + y^2)^{\frac{1}{2}} \quad \text{and} \quad v(x, y) = 0.$$

Then $v_x = v_y = 0$ and for $(x, y) \neq (0, 0)$,

$$u_x = x(x^2 + y^2)^{-\frac{1}{2}}, \qquad u_y = y(x^2 + y^2)^{-\frac{1}{2}}.$$

The Cauchy–Riemann equations fail to hold, and f to be differentiable, at any point $z \neq 0$. The point 0 requires separate attention. From first principles,

$$\frac{f(h) - f(0)}{h} = \frac{|h|}{h} \to \begin{cases} 1 & \text{as } h \to 0 \quad (h \text{ real and positive}), \\ -1 & \text{as } h \to 0 \quad (h \text{ real and negative}). \end{cases}$$

Hence $f'(0)$ does not exist. $\qquad \square$

2 The Cauchy–Riemann equations are useful for proving non-differentiability. They are *not*, on their own, a sufficient condition for differentiability (see Exercise 2.3). we return to this point in Chapter 10. There we also use the Cauchy–Riemann equations to establish that the real and imaginary parts of a holomorphic function satisfy Laplace's equation in two dimensions. This fact forms the basis of many of the applications of complex analysis to physical problems.

The Cauchy–Riemann equations also have theoretical applications, as the next result shows.

2.6 Proposition

Suppose $f \in H(D(0; R))$. Then
(1) if $f' = 0$ in $D(0; R)$, f is constant;
(2) if $|f|$ is a constant, c, in $D(0; R)$, f is constant.

Proof. We adopt the notation of Theorem 2.4. The proof of this theorem shows that for $z = x + iy \in D(0; R)$,

$$f'(z) = u_x + iv_x = -iu_y + v_y.$$

Suppose $f' = 0$. Then $u_x = v_x = u_y = v_y = 0$. Fix arbitrary points $p = a + ib$ and $q = c + id \in D(0; R)$. We shall show that $f(p) = f(q)$. At least one of $s = c + ib$ and $t = a + id$ lies in $D(0; R)$; without loss of generality suppose s does. Each of $x \mapsto u(x, b)$ and $y \mapsto u(c, y)$ is a real-valued function of a real variable with zero derivative and so is constant, by the Mean-value theorem. Hence

$$u(a, b) = u(c, b), \qquad u(c, b) = u(c, d),$$

and similarly

$$v(a, b) = v(c, b), \qquad v(c, b) = v(c, d).$$

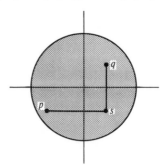

Fig. 2.1

We conclude that $f(p) = f(s) = f(q)$.

Now suppose that $|f| = c$, so that $u^2 + v^2 = c^2$. Then

$$uu_x + vv_x = 0, \qquad uu_y + vv_y = 0,$$

whence, by the Cauchy–Riemann equations,

$$uu_x - vu_y = 0, \qquad uu_y + vu_x = 0.$$

Elimination of u_y gives $0 = (u^2 + v^2)u_x = c^2 u_x$. If $c = 0$, f is trivially constant. Otherwise, $u_x = 0$, and similarly u_y, v_x, and v_y are zero. We deduce, as above, that f is constant. $\qquad\square$

This proof is unaesthetic, but instructive. It provides a stepping stone between complexified real analysis and complex analysis proper. It motivates the introduction, in Chapter 3, of connected sets; connectedness is the characteristic of $D(0; R)$ germane to the proof of Proposition 2.6; see Proposition 3.18.

2.7 Holomorphic functions: elementary properties and examples

We have so far refrained from giving examples of holomorphic functions because except in the simplest cases it is very laborious to check holomorphy from the definition. (As in real analysis, proving differentiability from first principles is useful mainly as an exercise in evaluating limits.) Instead, we build up a catalogue of holomorphic functions by forming products, composites, etc.

The following properties are stated for functions holomorphic in an arbitrary set S. They are proved by checking the appropriate differentiability conditions, pointwise on a suitable open set $G \supseteq S$ (see Definition 2.2(3)). We omit the details as the proofs are formally identical to their real counterparts.

(1) Let f and g be holomorphic in S and $\lambda \in \mathbb{C}$. Then λf, $f + g$, and fg (defined pointwise in the usual way) are holomorphic in S and the usual differentiation rules apply: for all $z \in S$,

$$(\lambda f)'(z) = \lambda f'(z), \qquad (f + g)'(z) = f'(z) + g'(z),$$
$$(fg)'(z) = f'(z)g(z) + f(z)g'(z).$$

(2) **The chain rule** Let f be holomorphic in S and let g be holomorphic in $f(S)$. Then the composite function $g \circ f$, given by $(g \circ f)(z) = g[f(z)]$, is holomorphic in S and, for all $z \in S$,

$$(g \circ f)'(z) = g'[f(z)]f'(z).$$

(3) Let f be holomorphic in S and suppose that for all $z \in S$, $f(z) \neq 0$. Then $1/f$ is holomorphic in S and

$$(1/f)'(z) = -f'(z)/f(z)^2.$$

We can now construct examples of holomorphic functions easily. The function f defined by $f(z) = z$ is obviously differentiable everywhere, as is any constant function. By (1), any *polynomial*

$$p(z) = \sum_{n=0}^{N} c_n z^n \quad (c_n \in \mathbb{C}, \ N \text{ an integer} \geq 0)$$

is holomorphic in $\mathbb{C}$. By (1) and (3), a *rational function* $p(z)/q(z)$ ($p(z)$, $q(z)$ polynomials) is holomorphic in any set in which $q(z)$ is never zero. Thus, for example, $(1 + z^2)^{-2}$ is holomorphic everywhere except at $\pm i$.

To enlarge our stock of holomorphic functions we progress from polynomials $\sum_{n=0}^{N} c_n z^n$, which are finite sums, to power series $\sum_{n=0}^{\infty} c_n z^n$ (or more generally $\sum_{n=0}^{\infty} c_n (z - a)^n$), which are infinite series raising questions of convergence.

Complex power series

2.8 Series of complex terms

Suppose $\langle a_n \rangle_{n \geq 0}$ is a complex sequence. The series $\sum a_n$ is said to *converge to the sum s* if the sequence $\langle s_n \rangle$ of partial sums, given by

$$s_n := a_0 + a_1 + \ldots + a_n,$$

converges to the limit s, in the sense of Definition 1.14. We write $s = \sum_{n=0}^{\infty} a_n$ (and this defines the expression on the right-hand side). Henceforth, where it would be pedantic to do otherwise, we do not distinguish between the series $\sum a_n$ and the sum, $\sum_{n=0}^{\infty} a_n$, to which it converges.

As an example we mention a series we shall use many times. For $\sum_{n=0}^{\infty} z^n$,

$$s_n = \frac{1 - z^{n+1}}{1 - z} \quad (z \neq 1),$$

so that for $|z| < 1$ the series converges to the sum $(1 - z)^{-1}$, while for $|z| \geq 1$ it fails to converge. Hence we have the binomial expansion

$$(1 - z)^{-1} = 1 + z + z^2 + \ldots \quad (|z| < 1).$$

Developing the theory of complex series is mainly a matter of checking that the same techniques work as in the real case. We collect together for reference some facts about complex series.

(1) Suppose $\sum a_n$ converges. Then $a_n \to 0$ as $n \to \infty$, and there exists a constant M such that $|a_n| \leq M$ for all n (cf. Binmore [4], 6.9 and 4.25).

(2) Suppose $\sum |a_n|$ converges. Then $\sum a_n$ converges. This result, which is expressed in words as 'absolute convergence implies convergence', can be deduced from its real analogue by considering $\operatorname{Re} a_n$ and $\operatorname{Im} a_n$. Alternatively, but essentially equivalently, one may prove it by first establishing the Cauchy criterion for convergence of a complex sequence, which can be obtained from Theorem 1.16 (cf. [4], 5.19).

(3) Suppose $\sum b_n$ is a convergent series with $b_n \geq 0$ for all n, and suppose that for some constant $k > 0$, $|a_n| \leq k b_n$ for all n. Then $\sum a_n$ converges. (This is a combination of (2) with the Comparison test for real series; cf. [4], 6.15.)

(4) Given any complex series $\sum a_n$, $\sum |a_n|$ is a series with real non-negative terms, to which well-known tests for convergence— d'Alembert's ratio test, Cauchy's root test, etc.—apply.

2.9   Definition

Given a *power series* $\sum c_n z^n$ ($c_n \in \mathbb{C}$ ($n \geq 0$)), the *radius of convergence*, R, is defined by

$$R := \sup\{|z| : \sum |c_n z^n| \text{ converges}\}$$

(here $R = \infty$ is allowed in the case that the series converges for arbitrarily large $|z|$).

Suppose $|z| < R$. Then, by definition of supremum, there exists w such that $|z| < |w| < R$ and $\sum |c_n w^n|$ converges. Since $|c_n z^n| \leq |c_n w^n|$, 2.8(3) implies that $\sum c_n z^n$ converges. The definition of R is phrased in terms of absolute convergence for ease of computation; see the note above and Example 2.10.

2.10   Example

Find the radius of convergence of (1) $\sum (-1)^n n^2 z^n$, and (2) $\sum z^n / n!$.

Solution. (1) Apply the Ratio test to $\sum |(-1)^n n^2 z^n|$. For $z \neq 0$,

$$\left| \frac{(-1)^{n+1}(n+1)^2 z^{n+1}}{(-)^n n^2 z^n} \right| = (1 + 1/n)^2 |z| \to |z| \quad \text{as} \quad n \to \infty.$$

Hence $\sum |(-1)^n n^2 z^n|$ converges if $|z| < 1$ and fails to converge if $|z| > 1$, so the radius of convergence R is 1.

(2) For $z \neq 0$,

$$\left| \frac{z^{n+1}/(n+1)!}{z^n/n!} \right| = \frac{|z|}{n+1} \to 0 \quad \text{as} \quad n \to \infty,$$

and the Ratio test implies that the power series converges absolutely. Hence $R = \infty$. $\qquad\square$

Our interest in power series is in their behaviour as functions. As already noted, a power series $\sum c_n z^n$ with non-zero radius of convergence R converges for $|z| < R$, and so we can define a function f by $f(z) = \sum_{n=0}^\infty c_n z^n$ ($|z| < R$). We shall establish that $f \in H(D(0; R))$. The proofs of Theorem 2.12 and the preceding lemma are unappealingly technical, but the theorem is important. The idea is to show that "term-by-term" differentiation is legitimate.

2.11 Lemma

The power series $\sum c_n z^n$ and $\sum n c_n z^{n-1}$ have the same radius of convergence.

Proof. We prove that if $\sum |c_n z^n|$ converges for $|z| < R$ ($\neq 0$), then so does $\sum |n c_n z^{n-1}|$, leaving the (easier) converse, which we do not need later, as an exercise.

Fix z with $0 < |z| < R$ and choose ρ so that $|z| < \rho < R$. Then

$$|n c_n z^{n-1}| = \frac{n}{|z|} \left(\frac{|z|}{\rho} \right)^n |c_n \rho^n|.$$

The series $\sum n(|z|/\rho)^n$ is easily shown to converge, by the Ratio test. Hence, by 2.8(1), there exists a constant M such that for all n,

$$n(|z|/\rho)^n \leqslant M,$$

whence

$$|n c_n z^{n-1}| \leqslant \frac{M}{|z|} |c_n \rho^n|,$$

from which the result follows by 2.8(3). $\qquad\square$

2.12 Theorem

Let $\sum c_n z^n$ have radius of convergence $R \neq 0$. Define f by $f(z) = \sum_{n=0}^\infty c_n z^n$. Then $f \in H(D(0; R))$ and

$$f'(z) = \sum_{n=1}^\infty n c_n z^{n-1} \qquad (|z| < R).$$

Proof. Lemma 2.11 implies that $g(z) = \sum_{n=1}^{\infty} nc_n z^{n-1}$ is well-defined for $|z| < R$. We shall prove that $f'(z)$ exists and equals $g(z)$. To do this we shall need the binomial expansion

$$(z+h)^n = \sum_{k=0}^{n} \binom{n}{k} z^{n-k} h^k;$$

this can be proved by induction just as in the real case. For z and $z+h$ in $D(0; R)$,

$$\left| \frac{f(z+h)-f(z)}{h} - g(z) \right| = \left| \sum_{n=1}^{\infty} c_n \left(\frac{(z+h)^n - z^n}{h} - nz^{n-1} \right) \right|$$

$$= |h| \left| \sum_{n=1}^{\infty} c_n \sum_{k=2}^{n} \binom{n}{k} z^{n-k} h^{k-2} \right|$$

$$\leq |h| \sum_{n=1}^{\infty} \tfrac{1}{2} n(n-1) |c_n| \sum_{m=0}^{n-2} \binom{n}{m} |z|^{n-2-m} |h|^m$$

$$\leq |h| \sum_{n=1}^{\infty} \tfrac{1}{2} n(n-1) |c_n| (|z|+|h|)^{n-2}.$$

Fix z and choose ρ with $|z| < \rho < R$. By Lemma 2.11, $\sum_{n=1}^{\infty} n(n-1) |c_n| \rho^{n-2}$ converges, to K say. For $|h| < \rho - |z|$,

$$\left| \frac{f(z+h)-f(z)}{h} - g(z) \right| \leq \tfrac{1}{2} K |h|.$$

Hence $f'(z)$ exists and equals $g(z)$. □

2.13 Corollary

Suppose $f(z) = \sum_{n=0}^{\infty} c_n z^n$, the power series having non-zero radius of convergence. Then f has derivatives of all orders at 0 and, for $n = 0, 1, 2, \ldots, f^{(n)}(0) = n! c_n$.

2.14 Remarks

(1) Continuity of functions defined by power series can be proved directly, or deduced from 2.12 and 2.3(1).
(2) Corollary 2.13 shows that the coefficients in a power series expansion are unique.
(3) The above arguments also apply to $\sum c_n(z - a)^n$, for any $a \in \mathbb{C}$.

Elementary functions

We shall assume that readers know the basic properties and the Maclaurin expansions of the real trigonometric, exponential, and logarithmic functions. For some this knowledge will be founded on a naive treatment relying on elementary geometry and trigonometry, while others will have seen such an approach superseded by a sounder method in which functions are *defined* by power series and their familiar properties derived from these series. For complex functions a geometric approach is no longer available, but power series serve admirably. We illustrate how with a detailed treatment of complex exponentials.

2.15 The exponential function

The power series $\sum z^n/n!$ has infinite radius of convergence (Example 2.10(2)). We define the exponential function by

$$e^z := \sum_{n=0}^{\infty} \frac{z^n}{n!} \quad (z \in \mathbb{C}).$$

Theorem

(1) e^z is holomorphic (and hence continuous) in $\mathbb{C}$, and for all z,

$$\frac{d}{dz} e^z = e^z;$$

(2) for all z and w in $\mathbb{C}$, $e^{z+w} = e^z e^w$;
(3) $e^z \neq 0$ for any $z \in \mathbb{C}$ and $e^z > 0$ if z is real;
(4) for $z \in \mathbb{C}$, $|e^z| = e^{\operatorname{Re} z}$, and when z is real, $|e^{iz}| = 1$.

Proof. (1) is immediate from Theorem 2.12 and Remark 2.3(1).

To prove (2) we consider, for fixed $\zeta \in \mathbb{C}$, the function $f(z) = e^z e^{\zeta - z}$ and differentiate to get (by (1) and the chain rule, 2.7(2)),

$$f'(z) = e^z e^{\zeta - z} - e^z e^{\zeta - z} = 0.$$

By Proposition 2.6(1), f has the constant value $f(0)$ in any $D(0; R)$ and hence in $\mathbb{C}$. Thus, for all z and ζ, $e^\zeta = e^z e^{\zeta - z}$. Substituting $w + z$ for ζ yields (2).

(3) is a consequence of $e^z e^{-z} = e^0 = 1$ and the fact that $e^z > 0$ when z is real and positive.

To prove (4) we observe that

$$\begin{aligned}
|e^z|^2 &= e^z \overline{e^z} &&\text{(by 1.3(3))} \\
&= e^z e^{\bar{z}} &&\text{(using the definition of } e^z \text{ and 1.3(4))} \\
&= e^{z + \bar{z}} &&\text{(by (2))} \\
&= e^{2 \operatorname{Re} z} \\
&= (e^{\operatorname{Re} z})^2 &&\text{(by (2)).}
\end{aligned}$$

Hence $|e^z| = e^{\mathrm{Re}\,z}$ (since both sides are real and positive). If z is real, $\mathrm{Re}(iz) = 0$, so $|e^{iz}| = e^0 = 1$. $\qquad\qquad\qquad\square$

2.16 Trigonometric and hyperbolic functions

The complex cosine and sine functions are defined by power series as follows:

$$\cos z := \sum_{n=0}^{\infty} (-1)^n \frac{z^{2n}}{(2n)!}, \qquad \sin z := \sum_{n=0}^{\infty} (-1)^n \frac{z^{2n+1}}{(2n+1)!}.$$

Both power series have infinite radius of convergence and so define functions holomorphic in $\mathbb{C}$, and we have, by 2.12,

$$\frac{d}{dz}\cos z = -\sin z, \qquad \frac{d}{dz}\sin z = \cos z.$$

From the series definitions we obtain the relations

$$e^{iz} = \cos z + i \sin z,$$

$$2\cos z = (e^{iz} + e^{-iz}), \qquad 2i\sin z = (e^{iz} - e^{-iz}).$$

The hyperbolic functions cosh and sinh are defined by

$$\cosh z := \tfrac{1}{2}(e^z + e^{-z}) = \cos iz, \qquad \sinh z := \tfrac{1}{2}(e^z - e^{-z}) = -i\sin iz.$$

Other trigonometric and hyperbolic functions are defined in the expected way: for example,

$$\tan z := \frac{\sin z}{\cos z}.$$

For z real, all these functions agree with their real counterparts. The addition formulae, etc., known to hold for real z are also valid for complex z. We shall see in 5.18 how identities between holomorphic functions f and g which are true on $\mathbb{R}$ (or a non-trivial subinterval of it) persist in the intersection of the sets where f and g are holomorphic (subject to some topological qualifications). In the meantime we note that trigonometric formulae can be proved directly from the power series (cf. 2.15), and we shall feel free to make use of such formulae in the complex case.

By contrast, inequalities valid for functions defined on $\mathbb{R}$ need not persist when the functions are extended to $\mathbb{C}$. For z real, $|\cos z| \leqslant 1$. However, it is not true that $|\cos z| \leqslant 1$ for all $z \in \mathbb{C}$: if $z = iy$ (y real),

$$|\cos iy| = |\cosh y| \to \infty \quad \text{as} \quad y \to \infty.$$

Similarly, $\sin z$ is unbounded in $\mathbb{C}$.

We have the following identities: for x and y real,

$$e^{x+iy} = e^x e^{iy} = e^x(\cos y + i \sin y),$$
$$\cos(x+iy) = \cos x \cos iy - \sin x \sin iy = \cos x \cosh y - i \sin x \sinh y,$$
$$\sin(x+iy) = \sin x \cos iy + \cos x \sin iy = \sin x \cosh y + i \cos x \sinh y.$$

If we put $x = 0$ and $y = \theta$ (θ real) in the first of these, the fundamental equation

$$e^{i\theta} = \cos \theta + i \sin \theta$$

reappears. (Note that $|e^{i\theta}| = 1$ is immediate from this, assuming, which we did not in the proof of 2.15(4), properties of cosine and sine on $\mathbb{R}$.)

2.17 Zeros and periodicity

We have already seen that $e^z \neq 0$ for any $z \in \mathbb{C}$. It follows easily from the equations above that

$$\cos z = 0 \Leftrightarrow z = \tfrac{1}{2}(2k+1)\pi \qquad (k \in \mathbb{Z}),$$
$$\sin z = 0 \Leftrightarrow z = k\pi \qquad (k \in \mathbb{Z}),$$
$$\cosh z = 0 \Leftrightarrow z = \tfrac{1}{2}(2k+1)\pi i \qquad (k \in \mathbb{Z}),$$
$$\sinh z = 0 \Leftrightarrow z = k\pi i \qquad (k \in \mathbb{Z}).$$

It is also important to note that

$$e^z = 1 \Leftrightarrow z = 2k\pi i \qquad (k \in \mathbb{Z}),$$
$$e^z = -1 \Leftrightarrow z = (2k+1)\pi i \qquad (k \in \mathbb{Z}).$$

Further, $e^{z+\alpha} = e^z$ for all $z \in \mathbb{C}$ if and only if $\alpha = 2k\pi i$ for some $k \in \mathbb{Z}$, while $\cos(z + \alpha) = \cos z$ and $\sin(z + \alpha) = \sin z$ each hold for all z if and only if $\alpha = 2k\pi$ for some $k \in \mathbb{Z}$.

2.18 Argument

The familiar fact that on $\mathbb{R}$ cosine and sine are periodic of period 2π (with, if one wishes to be formal, π as the smallest positive solution of the equation $\sin x = 0$), underlies all the assertions in 2.17. It is also this periodicity which is responsible for the non-uniqueness of θ in the representation $z = |z| e^{i\theta}$. We can now be more precise about this non-uniqueness problem. For each $z \in \mathbb{C}$, $z \neq 0$, we define the *argument* of z to be

$$[\arg z] := \{\theta \in \mathbb{R} : z = |z| e^{i\theta}\}.$$

The bracket notation $[\arg z]$ is designed to emphasize that the argument of z is a set of numbers, not a single number. In fact, $[\arg z]$ is an infinite set, of the form $\{\theta + 2k\pi : k \in \mathbb{Z}\}$, where θ is any fixed number such that $e^{i\theta} = z/|z|$. For example, $[\arg i] = \{(4k + 1)\pi/2 : k \in \mathbb{Z}\}$. For z and w not equal to zero.

$$[\arg zw] = \{\theta + \phi : \theta \in [\arg z], \phi \in [\arg w]\},$$

$$[\arg 1/z] = \{-\theta : \theta \in [\arg z]\}.$$

Those accustomed to a principal value determination $\text{Arg } z = \theta$, where $z = |z| e^{i\theta}$, $-\pi < \theta \leqslant \pi$ (or alternatively $0 \leqslant \theta < 2\pi$), may wonder why we have shunned this in favour of the more complicated $[\arg z]$. As we show in Chapter 4, the difficulty is that we cannot impose a restriction which determines θ uniquely (such as $-\pi < \theta \leqslant \pi$) and simultaneously allow z to move freely in $\mathbb{C}\backslash\{0\}$ with θ varying continuously (see Exercise 4.8). If, for example, z performs a complete anticlockwise circuit round the unit circle, θ increases by 2π and a jump discontinuity in $\text{Arg } z$ is inevitable.

2.19   Complex logarithms

One way to define the logarithm on $(0, \infty) \subseteq \mathbb{R}$ is as the function inverse to the exponential function: for each positive real number x there exists a unique real solution $t = \log_e x$ to the equation $e^t = x$. (Since we shall only work with logarithms to the base e, we shall henceforth drop the subscript and write $\log x$ in place of $\log_e x$.) In the complex case we seek solutions to the equation $e^w = z$.

Suppose $z \in \mathbb{C}$, $z \neq 0$. Put $z = e^w = e^{u+iv}$ (u, v real). Then

$$|z| = |e^u e^{iv}| = e^u \quad \text{(by 2.15(4))}$$

and

$$[\arg z] = \{v + 2k\pi : k \in \mathbb{Z}\}.$$

We have proved the important relation

$$e^w = z \quad \text{if and only if} \quad w = \log|z| + i\theta, \quad \text{where} \quad \theta \in [\arg z].$$

We accordingly define, for $z \neq 0$,

$$[\log z] = \{\log|z| + i\theta : \theta \in [\arg z]\}.$$

For example, $[\log 2] = \{\log 2 + 2k\pi i : k \in \mathbb{Z}\}$ and $[\log(-i)] = \{(4k - 1)\pi i/2 : k \in \mathbb{Z}\}$. The complex logarithm is not a bona fide function, but a multifunction. We have assigned to each $z \neq 0$ infinitely many values of the logarithm.

As with $[\arg z]$, selection, for each $z \neq 0$, of a value from $[\log z]$ does not go far towards solving the many-valuedness problem, since we cannot expect a continuous, let alone a holomorphic, function to emerge on $\mathbb{C}\backslash\{0\}$. We return to this difficulty in Chapters 4 and 6, when we have the machinery for coping with it.

2.20 Roots and powers

If n is a positive integer and $z \neq 0$, there exist n solutions to the equation $w^n = z$, given in terms of the polar representation $z = r e^{i\theta}$ by $w = r^{1/n} e^{i(\theta + 2k\pi)/n}$. Note in particular the determination of complex nth roots of ± 1:

$$w^n = 1 \Leftrightarrow w = e^{2k\pi i/n}, \qquad w^n = -1 \Leftrightarrow w = e^{(2k+1)\pi i/n}$$

$$(k = 0, 1, \ldots, n-1).$$

More generally, if α is a complex number, we define, for $z \neq 0$,

$$[z^{\alpha}] = \{e^{\alpha(\log|z| + i\theta)} : \theta \in [\arg z]\}.$$

Note that e^{α} (as defined in 2.15) is one member of $[e^{\alpha}]$.

Only when α is an integer n does $[z^{\alpha}]$ not produce multiple values: in this case $[z^{\alpha}]$ contains the single point z^n. When $\alpha = 1/n$ $(n = 2, 3, \ldots)$ $[z^{\alpha}]$ contains the values of the nth root obtained above.

Complex powers must be treated with circumspection. The formula

$$x^{\alpha} x^{\beta} = x^{\alpha + \beta} \qquad (x, \alpha, \beta \text{ real})$$

can be shown to have a complex analogue (in which the values of the multifunctions involved have to be appropriately selected), but

$$x_1^{\alpha} x_2^{\alpha} = (x_1 x_2)^{\alpha} \qquad (x_1, x_2, \alpha \text{ real})$$

has no universally valid complex generalization.

Exercises

1. Let $f(z) = \text{Re } z$, $g(z) = \bar{z}$ and $k(z) = |z|^2$. Prove that f and g are not differentiable at any point of $\mathbb{C}$ and that k is differentiable only at 0.

2. Prove that $(\text{Im } z)^2$ is not differentiable anywhere.

3. Prove that f defined by

$$f(z) = z^5/|z|^4 \qquad (z \neq 0), \quad f(0) = 0,$$

satisfies the Cauchy–Riemann equations at $z = 0$ but is not differentiable there.

4. Where are the following holomorphic:
   (i) $z^8 + 7z^5 - \pi z^2 + 1$, (ii) $e^z/[z(z-1)(z-2)]$, (iii) $(z^5 - 1)^{-1}$,
   (iv) $(1 + e^z)^{-2}$, (v) $\sin(1/z)$, (vi) $z|z|$?

5. Where do the following series define holomorphic functions:
   (i) $\sum_{n=1}^{\infty} (-1)^n z^n/n$, (ii) $\sum_{n=0}^{\infty} z^{5n}$, (iii) $\sum_{n=0}^{\infty} z^n/n^n$, (iv) $\sum_{n=0}^{\infty} n^n z^n$?

6. Prove that, if a_n $(n = 0, 1, \ldots)$ are complex numbers such that $\sum |a_n|$ converges, then

$$\left| \sum_{n=0}^{\infty} a_n \right| \leqslant \sum_{n=0}^{\infty} |a_n|.$$

Deduce that, for all $z \in \bar{D}(0; 1)$,

$$(3-e)|z| \leqslant |e^z - 1| \leqslant (e-1)|z|.$$

7. Determine for which values of z the following series converge absolutely:

(i) $\displaystyle\sum_{n=0}^{\infty} \frac{(z+1)^n}{2^n}$,

(ii) $\displaystyle\sum_{n=0}^{\infty} \left(\frac{z-1}{z+1}\right)^n$,

(iii) $\displaystyle\sum_{n=1}^{\infty} \frac{1}{n^2}(z^n + z^{-n})$,

(iv) $\displaystyle\sum_{n=0}^{\infty} \frac{z^n}{1-z^n}$.

(Hint: remember that $a_n \to 0$ if $\sum |a_n|$ converges.)

8. Write down an expansion of the form $\sum_{n=0}^{\infty} c_n z^n$ for

(i) $\displaystyle\frac{1}{2z+5}$,

(ii) $\displaystyle\frac{1}{1+z^4}$,

(iii) $\displaystyle\frac{1+iz}{1-iz}$,

(iv) $\displaystyle\frac{1}{1+z+z^2}$,

(v) $\displaystyle\frac{1}{(z+1)(z+2)}$.

In each case specify where the expansion is valid.

9. Find the real and imaginary parts of (i) e^{2z}, (ii) e^{z^2}, (iii) e^{e^z}.

10. Suppose that f is holomorphic in $D(0; 1)$.
(i) Prove that if $\operatorname{Re} f$ is constant, then f is constant.
(ii) Prove that if e^f is constant, then f is constant.

11. For each of the following functions f find $\{z \in \mathbb{C} : f(z) = 0\}$:
(i) $(z^4 - 1)\sin \pi z$,
(ii) $\cosh^2 z$,
(iii) $z^6 + 1$,

(iv) $1 - e^{z^2}$,
(v) $\sin^3 \dfrac{1}{z}$.

12. Find all solutions of (i) $e^z = -1 + i\sqrt{3}$, (ii) $\sin z = 100$, (iii) $\cos^2 z = 4$, (iv) $\tan z = i$.

13. Find the image under the mapping $z \mapsto e^z$ of (i) $\operatorname{Re} z = a$, (ii) $\operatorname{Im} z = b$, where a and b are real constants.

14. Find (i) $[\arg(-1)]$, $[\arg(1 - i)]$, and $[\arg e^{-2\pi i/3}]$,
(ii) $[\log(-1)]$, $[\log(1 + i)]$, and $[\log e^z]$ (for $z \in \mathbb{C}$).

15. Find $[1^{\frac{1}{3}}]$, $[(-8)^{\frac{1}{3}}]$, $[i^{\frac{1}{2}}]$, and $[(-1)^{\frac{1}{4}}]$, and in each case plot the set as a subset of the complex plane.

16. Find $[\sqrt{2}^i]$, $[i^{\sqrt{2}}]$, $[i^i]$, and $[e^{i\pi}]$.

3 Prelude to Cauchy's theorem

Cauchy's theorem is the cornerstone of complex analysis and merits a chapter to itself. To keep Chapter 4 uncluttered with new definitions and preparatory theory, we use this chapter to assemble the necessary ingredients. These include the descriptions of paths of various types, the elements of integration in the complex plane, and the topological concepts of connectedness and simple connectedness. This list of prerequisites suggests that one needs a deep understanding of the geometry and topology of the plane to appreciate Cauchy's theorem to the full. Since such understanding cannot be instantly acquired, we have engineered a two-tier structure for our treatment of Cauchy's theorem. On Level I our goal is a form of the theorem designed for applications. The geometry is kept as simple as possible so that all the claims we make are intuitively reasonable, and can be substantiated using a minimum of topological machinery. However, merely to present a utilitarian form of Cauchy's theorem would be demeaning to complex analysis. We therefore give another version of the theorem, on Level II, involving simply connected regions. This offers greater topological insights and has some practical benefits too.

Paths

To regard a curve, such as a circle, simply as a subset of the plane will not suffice for our purposes. Instead we adopt a more dynamic approach and think of a curve as the route traced out by a moving point, the route being specified by a suitable function of some real parameter. For example, $\gamma(t) = e^{it}$ travels anticlockwise once round the unit circle as t increases from 0 to 2π. We now present the formal definitions. These are illustrated in Fig. 3.1 below.

3.1 Curves and paths
Let $[\alpha, \beta]$ $(-\infty < \alpha < \beta < \infty)$ be a compact interval in $\mathbb{R}$. A *curve* γ *with parameter interval* $[\alpha, \beta]$ is a continuous function $\gamma : [\alpha, \beta] \to$

C. It has *initial point* $\gamma(\alpha)$ and *final point* $\gamma(\beta)$, and is *closed* if $\gamma(\alpha) = \gamma(\beta)$. It is *simple* if $\alpha \le s < t \le \beta$ implies $\gamma(s) \ne \gamma(t)$ for $t - s < \beta - \alpha$.

Suppose γ is a curve with parameter interval $[\alpha, \beta]$. The image set $\gamma([\alpha, \beta]) := \{\gamma(t): t \in [\alpha, \beta]\}$ is denoted by γ^*. As the continuous image of a compact interval, γ^* is a compact subset of $\mathbb{C}$ (by 1.18). It carries a built-in orientation, determined by the direction in which $\gamma(t)$ traces out γ^* as t increases from α to β. Given γ, there exists a curve $-\gamma$ with the same image set but the opposite orientation:

$$(-\gamma)(t) := \gamma(\alpha + \beta - t) \qquad (t \in [\alpha, \beta]).$$

Let $\alpha \le \alpha_1 < \beta_1 \le \beta$. By restricting the function γ to $[\alpha_1, \beta_1]$, we obtain a new curve, which we denote by $\gamma \upharpoonright [\alpha_1, \beta_1]$. Now suppose $\alpha < \tau < \beta$ and let $\gamma_1 = \gamma \upharpoonright [\alpha, \tau]$ and $\gamma_2 = \gamma \upharpoonright [\tau, \beta]$. The final point of γ_1 coincides with the initial point of γ_2 (each is $\gamma(\tau)$), and γ^* is traced by first tracing γ_1^* and then tracing γ_2^*. Conversely, take curves γ_1 and γ_2 with parameter intervals $[\alpha_1, \beta_1]$ and $[\alpha_2, \beta_2]$. So long as $\gamma_1(\beta_1) = \gamma_2(\alpha_2)$, we can form the *join*, γ say, of γ_1 and γ_2. The recipe is

$$\gamma(t) := \begin{cases} \gamma_1(t) & \text{if } t \in [\alpha_1, \beta_1], \\ \gamma_2(t + \alpha_2 - \beta_1) & \text{if } t \in [\beta_1, \beta_1 + \beta_2 - \alpha_2]. \end{cases}$$

To avoid irritating technicalities later, the parameter intervals of γ_1 and γ_2 are here allowed to be arbitrary (whereas those of subcurves obtained by restriction automatically slot together). The penalty is a slightly complicated formula for join—essentially the parameter interval of γ_2 has to be translated. The joining process can be iterated: the join of $\gamma_1, \gamma_2, \ldots, \gamma_n$ can be defined provided the final point of γ_k coincides with the initial point of γ_{k+1} ($k = 1, \ldots, n - 1$).

A curve γ is said to be *smooth* if the function γ has a continuous derivative on its parameter interval $[\alpha, \beta]$. Here γ is, of course, a complex-valued function of a real variable; differentiability is defined exactly as for a real-valued function on $[\alpha, \beta]$, with the derivatives at α and β one-sided derivatives. Note also that, by Lemma 1.15(2), $\gamma'(t) = (\text{Re } \gamma)'(t) + (t) + i(\text{Im } \gamma)'(t)$ for each $t \in [\alpha, \beta]$. A *path* is the join of finitely many smooth curves.

Yet one more definition: a curve γ is said to *lie in a set S* if $\gamma^* \subseteq S$.

In the following illustrative diagrams we perforce depict γ^* (the image) rather than γ (the function). Arrows indicate the direction in which γ^* is traced. It should be noted that, even when γ is a

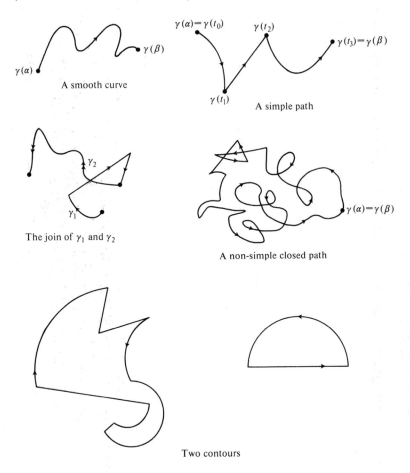

Fig. 3.1

path, γ^* may be geometrically extremely complicated, to an extent that diagrams cannot adequately convey.

3.2 Contours

Familiar figures such as circles and squares can be realized as images of paths. In particular

(A) for any u and v in $\mathbb{C}$, the image of the path γ given by

$$\gamma(t) = (1-t)u + tv \qquad (t \in [0, 1])$$

is the line segment $[u, v]$, traced from u to v, and

(B) any circular arc traced anticlockwise (clockwise) is the image of a path γ ($-\gamma$) where

$$\gamma(t) = a + re^{it} \qquad (t \in [\theta_1, \theta_2]),$$

for some $a \in \mathbb{C}$, $r > 0$, and $0 \leqslant \theta_1 < \theta_2 \leqslant 2\pi$.

We define a *contour* to be a simple closed path which is a finite join of paths each of type (A) or type (B). The image of a contour consists of finitely many line segments and circular arcs, and does not cross itself.

A geometric adjective (circular, triangular, . . .) applied to a path or contour will refer to the shape of its image. However we shall shorten 'triangular contour' to 'triangle', etc., where such an abuse of terminology will not cause confusion. We adopt the following notation. $\gamma(a; r)$ denotes the circle centre a radius r given by

$$\gamma(a; r)(t) := a + re^{it} \qquad (t \in [0, 2\pi]);$$
$$\Gamma_r(t) := re^{it} \qquad (t \in [0, \pi])$$

defines a frequently used semicircular arc, and $[u, v]$ denotes the line segment path in (A) above, as well as its image.

The reader should be warned that the term contour is customarily used in a wider sense. As we have defined them, contours encompass all the paths regularly arising in applied complex analysis, and have the virtue that their images are geometrically much simpler than those of arbitrary closed paths. Even for contours, the geometric properties we require—the existence of an 'inside' and an 'outside' for example—though obvious in most specific cases, are tricky to prove in general. For convenience, all proofs of results of this kind are deferred to the final section of this chapter.

Integration along paths

3.3 Integration of real- and complex-valued functions

We assume familiarity with integration of real-valued functions on compact intervals in $\mathbb{R}$, at least at a fairly basic level. Specifically, we take for granted simple techniques for evaluating real integrals, elementary properties of integrable functions, and the fact that, at the least, continuous functions on compact intervals are integrable. Here the reader may interpret integrable as having the meaning to which he is accustomed, whether he is acquainted with a Riemann-type approach or with Lebesgue integration.

Continuous functions will not be quite adequate for our needs. We say that a (real- or complex-valued) function h is *piecewise continuous* on a compact interval $[\alpha, \beta]$ of $\mathbb{R}$ if there exist points $\alpha = t_0 < t_1 < \ldots < t_n = \beta$ and continuous functions h_k on $[t_k, t_{k+1}]$ such that $h(t) = h_k(t)$ for $t \in (t_k, t_{k+1})$ $(k = 0, \ldots, n-1)$; h need not be defined at any or at some of the points t_k. Essentially the definition means that h is continuous apart from a finite number of jump discontinuities. A real-valued piecewise continuous function is integrable (in either the Riemann or Lebesgue sense); any reader not aware of this fact may, given h as above, take as a definition

$$\int_\alpha^\beta h(t)\, dt = \sum_{k=0}^{n-1} \int_{t_k}^{t_{k+1}} h_k(t)\, dt.$$

A complex-valued function g (defined on $[\alpha, \beta] \subseteq \mathbb{R}$) can be expressed as $g = \operatorname{Re} g + i \operatorname{Im} g$, where $\operatorname{Re} g$ and $\operatorname{Im} g$ are real-valued functions. We say g is *integrable* if and only if each of $\operatorname{Re} g$ and $\operatorname{Im} g$ is integrable, and in that case we define

$$\int g := \int \operatorname{Re} g + i \int \operatorname{Im} g.$$

For example,

$$\int_0^{2\pi} e^{it}\, dt = \int_0^{2\pi} \cos t\, dt + i \int_0^{2\pi} \sin t\, dt = \Big[\sin t\Big]_0^{2\pi} + i\Big[-\cos t\Big]_0^{2\pi} = 0.$$

In the sequel, where we manipulate complex-valued integrals without comment, we are using properties which carry over easily from the real-valued case; the linearity property $(\int (a g_1 + b g_2) = a \int g_1 + b \int g_2$ for a and $b \in \mathbb{C}$, with g_1 and g_2 complex-valued integrable functions) and the theorem on substitution used in Lemma 3.5 are examples. These properties can be found, for the real case, in Binmore [4] or Weir [8].

3.4 The integral of a function along a path

Let γ be a path with parameter interval $[\alpha, \beta]$. There exist points $\alpha = t_0 < t_1 < \ldots < t_n = \beta$ such that γ restricted to each $[t_k, t_{k+1}]$ coincides with a continuously differentiable function on $[t_k, t_{k+1}]$. At the points t_k, γ' need not exist. Let $f : \gamma^* \to \mathbb{C}$ be continuous. Then $(f \circ \gamma)\gamma'$ is piecewise continuous, and hence integrable, on $[\alpha, \beta]$. We define

$$\int_\gamma f(z)\, dz := \int_\alpha^\beta f[\gamma(t)]\gamma'(t)\, dt,$$

and call this the *integral of f along* γ, or *round* γ if γ is closed. Motivation for this definition is provided by the purely formal substitution of $\gamma(t)$ for z and $\gamma'(t)\,dt$ for dz. Before we turn to examples we clear up some technical points.

3.5 Lemma

Suppose that γ is a path with parameter interval $[\alpha, \beta]$ and that $f: \gamma^* \to \mathbb{C}$ is continuous.

(1) $\int_{-\gamma} f(z)\,dz = -\int_{\gamma} f(z)\,dz.$

(2) Let $\alpha < \tau < \beta$ and let $\gamma_1 = \gamma\!\restriction\![\alpha, \tau]$ and $\gamma_2 = \gamma\!\restriction\![\tau, \beta]$. Then

$$\int_{\gamma} f(z)\,dz = \int_{\gamma_1} f(z)\,dz + \int_{\gamma_2} f(z)\,dz.$$

(3) **Reparametrization** Let $\tilde{\gamma} = \gamma \circ \psi$, where ψ is a function mapping the parameter interval $[\tilde{\alpha}, \tilde{\beta}]$ of $\tilde{\gamma}$ onto $[\alpha, \beta]$ and having a positive continuous derivative. Then

$$\int_{\tilde{\gamma}} f(z)\,dz = \int_{\gamma} f(z)\,dz.$$

Proof. (1) and (2) are easily deduced from the definitions. In proving (3), we may, because of (2), assume γ, and hence also $\tilde{\gamma}$, to be smooth. For $t \in [\tilde{\alpha}, \tilde{\beta}]$,

$$\tilde{\gamma}'(t) = \gamma'[\psi(t)]\psi'(t)$$

(this is the chain rule in a real/complex hybrid form). Making the substitution $s = \psi(t)$, which is legitimate given the hypotheses on ψ, we have

$$\int_{\tilde{\gamma}} f(z)\,dz = \int_{\tilde{\alpha}}^{\tilde{\beta}} f[\tilde{\gamma}(t)]\tilde{\gamma}'(t)\,dt$$

$$= \int_{\tilde{\alpha}}^{\tilde{\beta}} f\{\gamma[\psi(t)]\}\gamma'[\psi(t)]\psi'(t)\,dt$$

$$= \int_{\alpha}^{\beta} f[\gamma(s)]\gamma'(s)\,ds$$

$$= \int_{\gamma} f(z)\,dz. \qquad \square$$

3.6 Corollary

Let γ be the join of paths $\gamma_1, \gamma_2, \ldots, \gamma_n$ and suppose $f : \gamma^* \to \mathbb{C}$ is continuous. Then

$$\int_\gamma f(z)\,dz = \sum_{j=1}^n \int_{\gamma_j} f(z)\,dz.$$

Lemma 3.5(3) tells us that under quite mild conditions $\int_\gamma f(z)\,dz$ depends on γ^* and the direction in which it is traced, but not on the parametrization chosen. It shows in particular that the parameter interval of a path γ can be translated and rescaled without $\int_\gamma f(z)\,dz$ being affected. Such a translation is needed in the proof of the corollary; recall the definition of join in 3.1.

3.7 Examples

(1) To compute $\int_\gamma (z-a)^n\,dz$, where γ is the circle $\gamma(a; r)$ and n is an integer.

Solution. Since $\gamma(t) = a + re^{it}$ $(t \in [0, 2\pi])$, the definition in 3.4 gives

$$\int_\gamma (z-a)^n\,dz = \int_0^{2\pi} (re^{it})^n ire^{it}\,dt = ir^{n+1} \int_0^{2\pi} e^{i(n+1)t}\,dt$$

$$= \begin{cases} 0 & (n \neq -1) \\ 2\pi i & (n = -1) \end{cases}$$

(as in the example in 3.3). $\qquad\square$

(2) To compute $\int_\gamma z^2\,dz$, where γ is the semicircular contour formed by joining $\gamma_1 := [-R, R]$ and $\gamma_2 := \Gamma_R$; see Fig. 3.2.

Solution. $\gamma_1(t) = (1-t)(-R) + tR,$ $\qquad \gamma_1'(t) = 2R$ $\qquad (t \in [0, 1])$
and
$$\gamma_2(t) = Re^{it}, \qquad \gamma_2'(t) = iRe^{it} \qquad (t \in [0, \pi]).$$

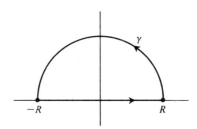

Fig. 3.2

Hence, using Corollary 3.6,

$$\int_\gamma z^2 \, dz = \int_0^1 [(2t-1)R]^2 2R \, dt + \int_0^\pi R^2 e^{2it} iRe^{it} \, dt$$

$$= \left[2R^3(\tfrac{4}{3}t^3 - 2t^2 + t) \right]_0^1 + \left[\tfrac{1}{3}R^3 e^{3it} \right]_0^\pi$$

$$= 0. \qquad \qquad \square$$

The usual way to evaluate real definite integrals is to recognize the integrand as a (continuous) derivative and then to apply the Fundamental theorem of calculus. There is an analogous result for complex integrals.

3.8 The Fundamental theorem of calculus

Suppose that γ is a path with parameter interval $[\alpha, \beta]$, that F is defined on an open set containing γ^*, and that $F'(z)$ exists and is continuous at each point of γ^*. Then

$$\int_\gamma F'(z) \, dz = \begin{cases} F[\gamma(\beta)] - F[\gamma(\alpha)] & \text{in general,} \\ 0 & \text{if } \gamma \text{ is closed.} \end{cases}$$

Proof. We first assume that γ is smooth. The hypotheses on F are more than strong enough to imply that $F \circ \gamma$ is differentiable on $[\alpha, \beta]$, with $(F \circ \gamma)'(t) = F'[\gamma(t)]\gamma'(t)$. Then

$$\int_\gamma F'(z) \, dz = \int_\alpha^\beta F'[\gamma(t)]\gamma'(t) \, dt$$

$$= \int_\alpha^\beta (F \circ \gamma)'(t) \, dt$$

$$= \int_\alpha^\beta \text{Re}\,(F \circ \gamma)'(t) \, dt + i \int_\alpha^\beta \text{Im}\,(F \circ \gamma)'(t) \, dt$$

$$= \left[\text{Re}(F \circ \gamma)(t) + i \,\text{Im}\,(F \circ \gamma)(t) \right]_\alpha^\beta$$

$$= F[\gamma(\beta)] - F[\gamma(\alpha)].$$

The penultimate line is obtained by applying the real Fundamental theorem of calculus ([4], 13.14, or [8], p. 57) to $\text{Re}\,(F \circ \gamma)$ and $\text{Im}\,(F \circ \gamma)$.

In the general case choose $\alpha = t_0 < t_1 < \ldots < t_n = \beta$ such that, for

$k = 0, \ldots, n-1$, $\gamma \restriction [t_k, t_{k+1}]$ is smooth. By the above,

$$\int_\gamma F'(z)\, dz = \sum_{k=0}^{n-1} \int_{t_k}^{t_{k+1}} F'[\gamma(t)]\gamma'(t)\, dt$$

$$= \sum_{k=0}^{n-1} \{F[\gamma(t_{k+1})] - F[\gamma(t_k)]\}$$

$$= F[\gamma(\beta)] - F[\gamma(\alpha)]. \qquad \square$$

Note that the calculations in Examples 3.7 are special cases of those in the above proof. The Fundamental theorem of calculus in complex analysis should, in a way that its real counterpart is not, be regarded as an interim result. It is a stepping stone to Cauchy's theorem and consequences thereof, and these largely supersede it for computational purposes.

When complex integrals cannot be evaluated explicitly (and sometimes where they can) the following estimate of magnitude is invaluable.

3.9 The Estimation theorem

Suppose that γ is a path with parameter interval $[\alpha, \beta]$ and that $f : \gamma^* \to \mathbb{C}$ is continuous. Then

$$\left| \int_\gamma f(z)\, dz \right| \leq \int_\alpha^\beta |f[\gamma(t)]\gamma'(t)|\, dt.$$

Proof. For a real-valued integrable function g on $[\alpha, \beta]$,

$$\left| \int_\alpha^\beta g(t)\, dt \right| \leq \int_\alpha^\beta |g(t)|\, dt.$$

Now

$$\left| \int_\gamma f(z)\, dz \right| = \left| \int_\alpha^\beta f[\gamma(t)]\gamma'(t)\, dt \right| = e^{i\phi} \int_\alpha^\beta f[\gamma(t)]\gamma'(t)\, dt,$$

for some real number ϕ, so that

$$\left| \int_\gamma f(z)\, dz \right| = \left| \int_\alpha^\beta \mathrm{Re}\, \{e^{i\phi} f[\gamma(t)]\gamma'(t)\}\, dt \right|$$

(because the left hand side is real and non-negative). Apply the inequality cited above to $g(t) := \mathrm{Re}\, \{e^{i\phi} f[\gamma(t)]\gamma'(t)\}$. It gives

$$\left| \int_\gamma f(z)\, dz \right| \leq \int_\alpha^\beta |\mathrm{Re}\, \{e^{i\phi} f[\gamma(t)]\gamma'(t)\}|\, dt,$$

whence the result follows, by 1.4(1). $\qquad \square$

3.10 Corollary

Let γ be a path with parameter interval $[\alpha, \beta]$ and let $f: \gamma^* \to \mathbb{C}$ be a continuous function satisfying $|f(z)| \leqslant M$ for all $z \in \gamma^*$. Then

$$\left| \int_\gamma f(z)\, dz \right| \leqslant M \int_\alpha^\beta |\gamma'(t)|\, dt.$$

We let length $(\gamma) := \int_\alpha^\beta |\gamma'(t)|\, dt$. For line segments and circular arcs, and hence also for contours, this definition gives the value expected for the length.

3.11 Examples

(1) Let $f(z) = (z^4 + 1)^{-1}$ and let $\gamma = \Gamma_R$. Then, by definition, $\int_\gamma f(z)\, dz = \int_0^\pi (R^4 e^{4it} + 1)^{-1} i Re^{it}\, dt$, the value of which is not obvious. However we do have

$$\left| \int_\gamma f(z)\, dz \right| \leqslant \int_0^\pi \left| \frac{Rie^{it}}{R^4 e^{4it} + 1} \right|\, dt \leqslant \frac{R\pi}{|R^4 - 1|} \quad \text{(by 3.10 and 1.4(3)).}$$

(2) Take $f(z) = 1/z$, $\gamma(t) = e^{it}$ $(t \in [0, 2\pi])$. Then $|f[\gamma(t)]| = 1$ and $|\gamma'(t)| = 1$. Then the Estimation theorem gives

$$\left| \int_\gamma f(z)\, dz \right| \leqslant \int_0^{2\pi} 1\, dt = 2\pi,$$

which is consistent with 3.7(1). Compare this with the fallacious estimate

$$\left| \int_\gamma f(z)\, dz \right| \leqslant \int_\gamma |f(z)|\, dz = \int_\gamma 1\, dz = \int_0^{2\pi} ie^{it}\, dt = 0.$$

The error lies in the fact the moduli must enclose the entire parametrized integrand $f[\gamma(t)]\gamma'(t)$ and not just $f[\gamma(t)]$ $(=f(z))$. A legitimate shorthand for $\int |f[\gamma(t)]\gamma'(t)|\, dt$ is $\int_\gamma |f(z)|\, |dz|$, which must not be confused with $\int_\gamma |f(z)|\, dz$.

3.12 Interchange of summation and integration

Suppose γ is a path and $u_0, u_1, \ldots$ are continuous functions on γ^*. It is certainly true that for any natural number N,

$$\sum_{k=0}^N \int_\gamma u_k(z)\, dz = \int_\gamma \sum_{k=0}^N u_k(z)\, dz.$$

If the finite sum here is replaced by an infinite sum, the corresponding interchange of summation and integration may well not be valid. A systematic study of when it is would require us to introduce, or assume known, basic facts about uniform convergence, or techniques more sophisticated still. Since we need only to be able to handle the integration of series closely related to power series, taken round circular contours, we elect to avoid uniform convergence *per se* and opt for a more *ad hoc* approach.

[Those who do know about uniform convergence will recognize Weierstrass's M-test lurking in Theorem 3.13. They should also realize that our hypotheses are stronger than necessary, and that the continuity of U is automatic from the other conditions.]

3.13 Theorem

Suppose that γ is a path, that $U, u_0, u_1, \ldots$ are continuous on γ^*, and that for all $z \in \gamma^*$, $\sum_{k=0}^{\infty} u_k(z)$ converges to $U(z)$. Suppose further that there exist constants M_k such that $\sum M_k$ converges and, for all $z \in \gamma^*$, $|u_k(z)| \leq M_k$. Then

$$\sum_{k=0}^{\infty} \int_\gamma u_k(z)\,dz = \int_\gamma \sum_{k=0}^{\infty} u_k(z)\,dz = \int_\gamma U(z)\,dz.$$

Proof. For $n = 0, 1, \ldots$, let $U_n(z) = \sum_{k=0}^{n} u_k(z)$. Both U_n and U are continuous, and hence integrable, on γ^*. Also, by 2.8(3), $\sum |u_k(z)|$ converges. We now have

$$\left| \int_\gamma U(z)\,dz - \sum_{k=0}^{n} \int_\gamma u_k(z)\,dz \right|$$

$$= \left| \int_\gamma (U(z) - U_n(z))\,dz \right|$$

$$\leq \sup_{z \in \gamma^*} |U(z) - U_n(z)| \times \text{length}(\gamma) \quad \text{(by 3.10)}$$

$$\leq \sup_{z \in \gamma^*} \sum_{k=n+1}^{\infty} |u_k(z)| \times \text{length}(\gamma) \quad \text{(see Exercise 2.5)}$$

$$\leq \sum_{k=n+1}^{\infty} M_k \times \text{length}(\gamma),$$

and this tends to zero as $n \to \infty$, because $\sum M_k$ converges. $\square$

For an illustration of the use of this theorem, see, for example, the proof of Theorem 5.9.

Connectedness and simple connectedness

This section deals with some important topological ideas in which paths play a role. The Cauchy–Riemann equations show that a function f which is holomorphic in an open disc D, and such that $f' = 0$ in D, is necessarily a constant (see Proposition 2.6). On the other hand, if we let $\hat{D}$ be the union of two non-overlapping open discs, say $\hat{D} = D(2; 1) \cup D(-2; 1)$, then we can find a non-constant holomorphic function with zero derivative: define g by

$$g(z) = \begin{cases} 1 & \text{if} \quad z \in D(2; 1), \\ -1 & \text{if} \quad z \in D(-2; 1). \end{cases}$$

These contrasting situations are hardly surprising: the domain sets D and $\hat{D}$ are topologically quite different. Informally, $\hat{D}$ splits into two separate open subsets, while no such decomposition of D exists. Furthermore, any two points in D are the endpoints of a polygonal path (as defined in 3.2) lying in D (in fact of a single line segment), but no polygonal path with endpoints 2 and -2 lies entirely in $\hat{D}$.

3.14 Definitions

(1) A *region* is a non-empty open subset of $\mathbb{C}$ which is *connected*, that is, which cannot be expressed as $G_1 \cup G_2$, where G_1 and G_2 are non-empty, open, and disjoint.

(2) A subset G of $\mathbb{C}$ is *polygonally connected* if, given any two points a and b in G, there exists a polygonal path lying in G and having endpoints a and b.

As our earlier remarks suggest, these concepts are related.

3.15 Theorem

A non-empty open set G is a region if and only if it is polygonally connected.

Proof. Suppose G is a region and fix $a \in G$. Let

$$G_1 = \{z \in G : \text{there exists a polygonal path in } G \text{ with endpoints } a \text{ and } z\}$$

and let $G_2 = G \backslash G_1$. We require $G_1 = G$. We shall prove that each of G_1 and G_2 is open. Connectedness of G will then imply that one of these sets is empty. This cannot be G_1, since $a \in G_1$.

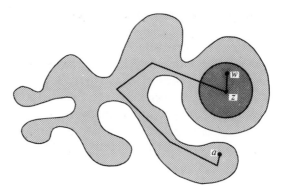

Fig. 3.3

We now establish our claim that G_1 and G_2 are open. For any $z \in G$, we can find r such that $D(z; r) \subseteq G$. For each $w \in D(z; r)$, $[z, w] \subseteq G$. It follows that z can be joined to a by a polygonal path in G if and only if w can be (see Fig. 3.3). Hence, for $k = 1, 2$, $z \in G_k$ implies $D(z; r) \subseteq G_k$.

Conversely, suppose G is a non-empty, open, polygonally connected set. Assume, for a contradiction, that G can be expressed as $G_1 \cup G_2$, where G_1 and G_2 are open, non-empty, and disjoint. Fix $a \in G_1$ and $b \in G_2$, and take a polygonal path γ in G with endpoints a and b. At least one constituent line segment of γ has one endpoint, p say, in G_1 and the other, q say, in G_2. Put $\tilde{\gamma}(t) = (1-t)p + tq$ $(t \in [0, 1])$. Define $h: [0, 1] \to \mathbb{R}$ by

$$h(t) = \begin{cases} 0 & \text{if} \quad \tilde{\gamma}(t) \in G_1, \\ 1 & \text{if} \quad \tilde{\gamma}(t) \in G_2. \end{cases}$$

Then $h(0) = 0$ and $h(1) = 1$, but h takes no value between 0 and 1. The required contradiction follows from 1.19 once we establish that h is continuous. To see this, note that for $|s - t|$ sufficiently small, $\tilde{\gamma}(s)$ and $\tilde{\gamma}(t)$ either both belong to G_1 or both to G_2 (because G_1 and G_2 are open and $\tilde{\gamma}$ is continuous). $\qquad \square$

3.16 Examples

An important class of examples of regions, which includes the open discs, is provided by the non-empty open convex sets. A subset S of $\mathbb{C}$ is said to be *convex* if, given a and $b \in S$, $[a, b] \subseteq S$. Any convex set is certainly polygonally connected.

| A convex region | A non-convex region | Not a region |

Fig. 3.4

3.17 Remark

Theorem 3.15 characterizes regions using polygonal paths. There are obvious possible variants on this. Essentially the same proof works if we use, for example,

(a) polygonal paths made up out of horizontal and vertical line segments, or

(b) paths whose images consist of finitely many line segments and circular arcs.

We can now extend Proposition 2.6.

3.18 Proposition

Suppose f is holomorphic in a region G and that $f' = 0$ in G. Then f is constant.

Proof. By Remark 3.17, any two points in G can be joined by a path consisting of horizontal and vertical line segments. The proof is now a mild complication of that of Proposition 2.6. □

Readers opting for the basic, Level I, approach to Cauchy's theorem should omit the remainder of this section.

3.19 Deformation and homotopy

We introduced polygonal connectedness as a means of distinguishing topologically between a single open disc and the disjoint union of open discs. We now seek to employ paths to distinguish an open disc D from an open annulus A. Informally, of course, the essential difference is that A has a hole in it but D does not, and a closed path in A whose image encircles the hole cannot be shrunk, within

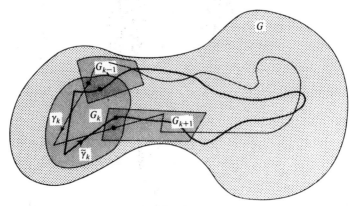

Fig. 3.5

A, to a point, while every closed path in *D* can be deformed to a point. We now make precise what we mean by deformation.

Let *G* be a non-empty open set in $\mathbb{C}$ and let γ and $\tilde{\gamma}$ be closed paths in *G*. We say that $\tilde{\gamma}$ can be obtained from γ by an *elementary deformation* if there exist open convex subsets G_0, G_1, ..., G_{N-1} of *G* such that γ may be expressed as the join of paths γ_0, γ_1, ..., γ_{N-1} and $\tilde{\gamma}$ as the join of paths $\tilde{\gamma}_0$, $\tilde{\gamma}_1$, ..., $\tilde{\gamma}_{N-1}$ in such a way that γ_k^* and $\tilde{\gamma}_k^* \subseteq G_k$ ($k = 0, \ldots, N-1$).

Further, closed paths γ and $\tilde{\gamma}$ in *G* are said to be *homotopic* (in *G*) if $\tilde{\gamma}$ can be obtained from γ by a finite number of elementary deformations.

Elementary deformation, which is illustrated in Fig. 3.5, is a more natural concept than it may seem from its somewhat ferocious definition. The idea is to cover the images of the two paths by overlapping convex regions and, within each of these, to replace a portion of γ^* by a portion of $\tilde{\gamma}^*$. The reason for working with convex sets emerges clearly in Chapter 4 (see the proof of 4.11).

Topologists have a definition of (closed path) homotopy, based on a continuous deformation process, which is, non-trivially, equivalent to our homotopy definition. The underlying idea is to take *G*, γ, and $\tilde{\gamma}$ as above and to think of a rubber band positioned over γ^*. The path $\tilde{\gamma}$ is homotopic to γ if the rubber band can be slid and stretched so as to coincide with $\tilde{\gamma}^*$ (correctly oriented), without ever moving outside *G*.

3.20 Definitions

A path γ lying in a set *G* is said to be *null* if $\gamma^* = \{a\}$ for some $a \in G$. A region *G* is *simply connected* if every closed path in *G* is homotopic to a null path in *G*.

3.21 Examples

(1) The definition of elementary deformation implies that any two closed paths in a convex region are homotopic. It follows that any convex region is simply connected. In particular, any disc $D(a; r)$ is simply connected.

(2) No open annulus is simply connected. This is eminently plausible, but non-trivial to prove. We give an indirect proof in 4.13.

(3) For any real number α, let

$$\mathbb{C}_\alpha := \mathbb{C} \setminus \{z \in \mathbb{C} : z = |z|\, e^{i\alpha}\}$$

(so that $\mathbb{C}_\alpha$ is the plane with a half-line emanating from 0 excluded). This region is simply connected. We leave the proof as a non-trivial exercise. See Supplementary Exercise 3.7.

Properties of paths and contours

The results in this section are needed later. Since all are obvious in special cases, and highly plausible in general, some readers may be content to omit the proofs. It is, however, instructive to see what is involved in proving the 'obvious'. In justifying geometric statements we have opted for an indication of strategy, at the expense of detail.

Our first goal is the Covering theorem, on which many later proofs rely. The theorem asserts that if γ is a path lying in an open set G, then its image γ^* can be covered by a finite chain of open discs contained in G, each overlapping the next. We give a proof with maximal geometric content. Our strategy is first to show that we can cover γ^* with discs all of the same radius, and then to show that only finitely many of these discs are needed. [Those with the requisite knowledge will appreciate that the Covering theorem is closely related to the Heine–Borel theorem.]

3.22 Lemma

Let γ be a path lying in an open set G. Then there exists a constant $m > 0$ such that for all $z \in \gamma^*$, $D(z; m) \subseteq G$.

Proof. We define

$$\rho(z) = \inf\{|z - w| : w \notin G\}$$

(so $\rho(z)$ is the distance of z from the complement of G). For any z and z' in $\mathbb{C}$, and w not in G,

$$\rho(z) \leqslant |z - w| \leqslant |z - z'| + |z' - w|,$$

whence $|z'-w| \geq \rho(z) - |z-z'|$. Taking the infimum over $w \notin G$, we get

$$\rho(z') \geq \rho(z) - |z-z'|.$$

Reversing the roles of z and z', we also have

$$\rho(z) \geq \rho(z') - |z-z'|.$$

These inequalities combine to give

$$|\rho(z) - \rho(z')| \leq |z-z'| \quad \text{for all } z \text{ and } z' \text{ in } \mathbb{C}.$$

We deduce that ρ is a continuous function. On the compact set γ^*, ρ is bounded and attains its infimum at some point $\zeta \in \gamma^*$ (by 1.18). Because G is open and $\zeta \in G$, there exists $m > 0$ such that $D(\zeta; m) \subseteq G$. Then, for all $z \in \gamma^*$, $\rho(z) \geq \rho(\zeta) \geq m$. By definition of ρ, $\rho(z) \geq m$ if and only if $D(z; m) \subseteq G$, so we have found a suitable constant m. $\qquad \Box$

3.23 The Covering theorem

Suppose γ is a path with parameter interval $[\alpha, \beta]$ such that γ^* is contained in G, where G is open. Then there exist a constant $m > 0$ and open discs $D_0, D_1, \ldots, D_N$ such that
   (i) for $k = 0, \ldots, N$, $D_k = D(\gamma(t_k); m)$, where $\alpha = t_0 < t_1 < \ldots < t_N = \beta$;
  (ii) for $k = 0, \ldots, N-1$, $D_k \cap D_{k+1} \neq \emptyset$;
 (iii) for $k = 0, \ldots, N-1$, $\gamma([t_k, t_{k+1}]) \subseteq D_k$;
 (iv) $\gamma^* \subseteq \bigcup_{k=0}^{N} D_k \subseteq G$.

Proof. Choose m as in Lemma 3.22, so that for every $z \in \gamma^*$, $D(z; m) \subseteq G$. It remains to show that γ^* may be covered by a *finite* chain of such discs, each overlapping the next, as in Fig. 3.6. If γ^* is made up of finitely many line segments and circular arcs (in particular if γ is a contour), this is clear from elementary geometry.

In the case of a general path we proceed as follows. Suppose first γ is smooth. Then the Mean-value theorem implies that for any s and t in $[\alpha, \beta]$,

$$(\operatorname{Re} \gamma)(s) - (\operatorname{Re} \gamma)(t) = (s-t)(\operatorname{Re} \gamma)'(c)$$

for some c between α and β, and similarly for $\operatorname{Im} \gamma$. The continuous functions $(\operatorname{Re} \gamma)'$, $(\operatorname{Im} \gamma)'$ are bounded on $[\alpha, \beta]$, by 1.18. It

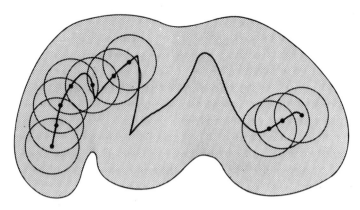

Fig. 3.6

follows that there exists $\delta > 0$ such that

$$|\gamma(s) - \gamma(t)| < m \quad \text{whenever} \quad |s - t| < \delta.$$

This conclusion [uniform continuity to the cognoscenti] persists for an arbitrary path, since we can apply the above argument to its constituent smooth curves.

We can now select points $\alpha = t_0 < t_1 < \ldots < t_N = \beta$ satisfying $|t_{k+1} - t_k| < \delta$ for $k = 0, 1, \ldots, N-1$. If, for $k = 0, \ldots, N$, D_k is chosen to be $D(\gamma(t_k); m)$, conditions (i)–(iv) of the theorem are met. $\qquad\square$

Remarks The disc D_N is not needed for the covering. It is put in for later notational convenience. When γ is closed, D_0 and D_N coincide.

We can actually do a little better than we have claimed in the theorem. It is possible to refine the argument so as to obtain, for some $\eta > 0$, an open strip $\{z : |z - w| < \eta \text{ for some } w \in \gamma^*\}$ containing γ^* and contained in G.

The famous Jordan curve theorem asserts that a simple closed path has an 'inside' and an 'outside'. In its general form, it is a very deep result. We restrict attention to contours (as defined in 3.2).

3.24 The Jordan curve theorem for a contour

Let γ be a contour. Then the complement of γ^* is of the form $I(\gamma) \cup O(\gamma)$, where $I(\gamma)$ and $O(\gamma)$ are disjoint connected open sets,

I(γ) (the *inside* of γ) is bounded and O(γ) (the *outside* of γ) is unbounded.

Outline proof. (For further details consult Kosniowski [6], pp. 102–103.) For any fixed $a \notin \gamma^*$, consider a half-line ℓ with endpoint a. Let $N(a, \ell)$ be the number of times ℓ cuts γ^* (this is well-defined except for, at worst, finitely many positions of ℓ involving tangency or 'corner points'; we leave $N(a, \ell)$ undefined in these degenerate cases). The crucial point to note is that whether $N(a, \ell)$ is odd or even depends only on a and not on the direction of ℓ; see Fig. 3.7. Let I(γ) (O(γ)) consist of those points $a \notin \gamma^*$ for which $N(a; \ell)$ is always odd (even).

That I(γ) and O(γ) are open follows from the observation that an open disc $D(z; r)$ disjoint from γ^* lies either wholly in I(γ) or wholly in O(γ) (given $w \in D(z; r)$, consider $[w, z]$ extended to a half-line). To prove connectedness of, say, I(γ), it is sufficient to show that any two points c and d in I(γ) can be joined by a path in I(γ) made up of line segments and circular arcs (see Remark 3.17). Figure 3.7 shows how this can be done: we join c and d to points c' and d' in I(γ) close to γ^* and, following γ^*, connect c' to d'. □

Our final result will allow us to break up an integral round a

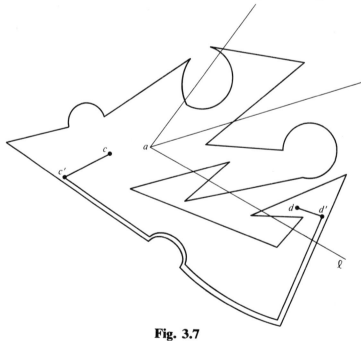

Fig. 3.7

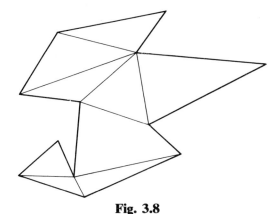

Fig. 3.8

polygonal contour into a sum of integrals round triangles, a crucial step in the proof of Cauchy's theorem (I) (4.6).

3.25 Triangulation of a polygon

Let γ be a polygonal contour in $\mathbb{C}$ and let $z_1, z_2, \ldots, z_n$ $(n > 3)$ be the vertices of γ^*. Then there exist $n - 3$ line segments $[z_j, z_k]$ each of which, excluding its endpoints, lies in $I(\gamma)$, and which are such that their insertion subdivides $I(\gamma)$ into $n - 2$ triangles.

Outline proof. (For further details, consult Hille [5], p. 286.) If $I(\gamma)$ is convex, then the segments $[z_1, z_k]$, $k = 3, \ldots, n - 1$, triangulate it. Otherwise, the interior angle at some vertex, say z_1, is greater than π. Consider a half-line ℓ emanating from z_1 such that $D'(z_1, r) \cap \ell \cap I(\gamma) \neq \varnothing$ for all r sufficiently small (so ℓ points into $I(\gamma)$). Moving along such a line from z_1, there is a first point of intersection w_ℓ ($\neq z_1$) of ℓ with γ^*. For at least one choice of ℓ, w_ℓ is a vertex. Let z_k be such a vertex. The segment $[z_1, z_k]$ can then be used to create two new polygonal contours, each of whose images in $\mathbb{C}$ has fewer than n vertices. The argument is repeated until only triangles remain. The process is illustrated in Fig. 3.8. $\square$

Exercises

1. Describe the image γ^* of the curve γ in the following cases.
   (i) $\gamma(t) = 1 + ie^{it}$ $(t \in [0, \pi])$.
   (ii) $\gamma(t) = e^{it}$ $(t \in [-\pi, 2\pi])$.
   (iii) γ is the join of $[-1, 1]$, $[1, 1+i]$, and $[1+i, -1-i]$.
   (iv) γ is the join of γ_1, γ_2, and γ_3, where γ_1 is $[1-i, 0]$, γ_2 is $[0, 1+i]$, and γ_3 is defined by $\gamma_3(t) = \sqrt{2}e^{i(t+\frac{1}{4}\pi)}$ $(t \in [0, 3\pi/2])$.
   (v) γ is given by $\gamma(t) = e^{it}$ $(t \in [0, \pi])$ and $\gamma(t) = e^{-it}$ $(t \in [\pi, 2\pi])$.
   (vi) $\gamma(t) = e^{it} \cos t$ $(t \in [0, 2\pi])$.
   In which cases is γ (a) closed, (b) simple, (c) a path, (d) smooth?

2. Define parametrically a path γ for which γ^* is
   (i) the square with vertices at $\pm 1 \pm i$,
   (ii) the closed semicircle in the right half-plane with $[-Ri, Ri]$ as diameter,
   (iii) the pair of circles $|z - 1| = 1$ and $|z + 1| = 1$, one traced clockwise, the other anticlockwise.

3. Evaluate $\int_\gamma f(z) \, dz$ when
   (i) $f(z) = |z|^4$, $\gamma = [-1 + i, 1 + i]$,
   (ii) $f(z) = z^2$, $\gamma(t) = e^{it} (t \in [-\pi/2, \pi/2])$,
   (iii) $f(z) = \text{Re } z$, $\gamma(t) = t + it^2 (t \in [0, 1])$,
   (iv) $f(z) = 1/z$, $\gamma(t) = e^{-it} (t \in [0, 8\pi])$,
   (v) $f(z) = e^z$, γ is the join of $[0, 1]$, $[1, 1 + i]$, and $[1 + i, i]$.

4. Evaluate $\int_\gamma |z|^4 \, dz$, $\int_\gamma (\text{Re } z)^2 \, dz$, $\int_\gamma z^{-2}(z^4 - 1) \, dz$, $\int_\gamma \sin z \, dz$ and $\int_\gamma z^{-1}(\bar{z} - \frac{1}{2}) \, dz$ when $\gamma = \gamma(0; 1)$. (Use 3.8 where it is applicable.)

5. Obtain an upper bound for $|\int_\gamma (z^2 + 4)^{-1} \, dz|$ when γ is (i) $[0, 1 + i]$, (ii) $\gamma(0; 10^3)$, (iii) $\gamma(i; 1)$.

6. By integrating $\dfrac{R + z}{z(R - z)}$ round a suitable contour, prove that

$$\frac{1}{2\pi} \int_0^{2\pi} \frac{R^2 - r^2}{R^2 - 2Rr \cos \theta + r^2} \, d\theta = 1 \quad (0 \leqslant r < R).$$

7. Suppose that $f(z) = \sum_{n=0}^\infty c_n z^n$ has radius of convergence $R > 0$. Use Theorem 3.13 to prove that

$$c_k = \frac{1}{2\pi i} \int_{\gamma(0;r)} \frac{f(z)}{z^{k+1}} \, dz \quad (0 \leqslant r < R).$$

Deduce that

$$r^k |c_k| \leqslant \sup\{|f(z)| : |z| = r\} \quad (0 \leqslant r < R).$$

8. Verify that the definition given in 3.10 for the length of a path γ gives the expected value when γ is a line segment or a circular arc.

9. (i) Suppose G_1 and G_2 are regions. Prove that $G_1 \cup G_2$ is a region if and only if $G_1 \cap G_2 \neq \varnothing$.
   (ii) Suppose G is a region and $z_1, \ldots, z_n$ are points in G. Prove that $G \setminus \{z_1, \ldots, z_n\}$ is a region.

10. Which of the sets described in Exercise 1.11 are (i) convex, (ii) regions, [(iii) simply connected regions]? (Give reasons, but not detailed formal proofs.)

11. A function f is holomorphic and real-valued in a region G. Prove that f is constant. Is this true if G is an arbitrary open set?

12. Let γ be a closed path and assume $z \notin \gamma^*$. By considering the function $w \mapsto |w - z|$ $(w \in \gamma^*)$ and appealing to 1.18(2), prove that there exists $\delta > 0$ such that $|w - z| \geqslant \delta$ for all $w \in \gamma^*$. Interpret this result geometrically.

4 Cauchy's theorem

Chapter 4 contains, as promised, various formulations of Cauchy's theorem. The conclusion of each is that $\int_\gamma f(z)\,dz = 0$ whenever $f \in H(G)$, where G is open and γ is a closed path in G; the differences lie in the further conditions imposed on γ and G to make this conclusion valid. Inextricably bound up with Cauchy's theorem are the Deformation theorem, concerning the equality of $\int_\gamma f(z)\,dz$ and $\int_{\tilde{\gamma}} f(z)\,dz$ under appropriate conditions, and the Antiderivative theorem, dealing with existence of $F \in H(G)$ such that $F' = F$. Our plan of campaign for proving these major theorems is outlined in the flow chart below.

Level I suffices for the multitude of applications in later chapters. The penultimate section of this chapter deals with logarithms, argument, and index. Those concentrating on Level I are recommended to postpone this section, but not to ignore it. The chapter

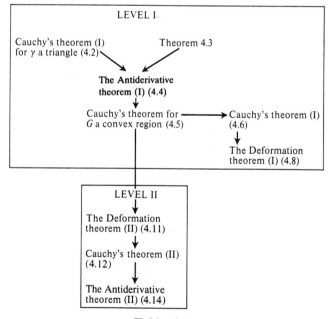

Table 4.1

concludes with a brief discussion, without proofs, of whether there is a definitive Cauchy theorem.

Cauchy's theorem, Level I

4.1 Note

Any contour γ (as defined in 3.2) has, by the Jordan curve theorem, 3.24, an inside $I(\gamma)$. Throughout this section, the hypothesis 'f is holomorphic inside and on γ' is to be taken to mean that f is holomorphic in $\gamma^* \cup I(\gamma)$, that is, that f is holomorphic in some open set containing $\gamma^* \cup I(\gamma)$ (see Definition 2.2(3)).

The Fundamental theorem of calculus given in 3.8 asserts that for a closed path γ in an open set G, $\int_\gamma F'(z)\,dz = 0$ for suitable functions F defined in G. A natural way to attack Cauchy's theorem is therefore to seek conditions under which $f \in H(G)$ has an antiderivative F (that is, $f = F'$). It turns out that, provided G is convex, a sufficient condition for this is: $\int_\gamma f(z)\,dz = 0$ for all triangles γ in G. Consequently we shall first prove that Cauchy's theorem is true in the special case that the path of integration is a triangle.

4.2 Cauchy's theorem for a triangle

Suppose that f is holomorphic inside and on a triangle γ. Then $\int_\gamma f(z)\,dz = 0$.

Proof. We first outline the ideas in the proof. The Fundamental theorem of calculus shows that $\int_{\tilde\gamma} p(z)\,dz = 0$ for any polynomial $p(z)$ and any triangular contour $\tilde\gamma$. Near a point Z, we can approximate our holomorphic function f by the polynomial $p(z) = f(Z) + (z - Z)f'(Z)$ (by 2.3(1)). Hence we aim to replace $\int_\gamma f(z)\,dz$ by the sum of integrals round small triangles on the image of which $p(z)$ is a good approximation to $f(z)$.

Let $[p, q, r]$ denote the triangle formed by joining $[p, q]$, $[q, r]$, and $[r, p]$. Let γ be $[u, v, w]$ and let u', v', and w' be, respectively, the midpoints of $[v, w]$, $[w, u]$, and $[u, v]$, as shown in Fig. 4.1. Consider the triangles $\gamma^0 = [u', v', w']$, $\gamma^1 = [u, w', v']$, $\gamma^2 = [v, u', w']$, and $\gamma^3[w, v', u']$. Then, by 3.5,

$$I := \int_\gamma f(z)\,dz = \sum_{k=0}^{3} \int_{\gamma^k} f(z)\,dz.$$

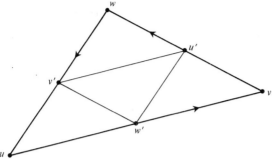

Fig. 4.1

For at least one value of k,

$$\left| \int_{\gamma^k} f(z)\, dz \right| \geq \tfrac{1}{4}|I| \quad \text{(by 1.4(3))}.$$

Relabel such a γ^k as γ_1. Repeat the argument with γ_1 in place of γ. Proceeding in this way, generate a sequence $\gamma_0, \gamma_1, \gamma_2, \ldots$ of triangles such that:

  (i) $\gamma_0 = \gamma$,
  (ii) for each n, $\Delta_{n+1} \subseteq \Delta_n$, where Δ_n is the closed triangular area having γ_n^* as its boundary,
  (iii) $\text{length}(\gamma_n) = 2^{-n} L$, where $L = \text{length}(\gamma)$, and
  (iv) $4^{-n}|I| \leq |\int_{\gamma_n} f(z)\, dz|$ for all $n \geq 0$.

The set $\bigcap_{n=0}^{\infty} \Delta_n$ contains a point Z common to all the triangles Δ_n. (To prove this, select for each n some point $z_n \in \Delta_n$. The sequence $\langle z_n \rangle$ is bounded since all points z_n belong to Δ_0. By Theorem 1.16, $\langle z_n \rangle$ has a subsequence convergent to some point Z. For each n, Z is a limit point of the subset $\{z_k : k \geq n\}$ of Δ_n and so belongs to Δ_n (see 1.11).)

Fix $\varepsilon > 0$. The function f is differentiable at Z, so, for some r,

$$|f(z) - f(Z) - (z - Z)f'(Z)| < \varepsilon |z - Z| \quad \text{for all } z \in D(Z; r). \quad (1)$$

Choose N such that $D(Z; r) \supseteq \Delta_N$. For such N,

$$|z - Z| \leq 2^{-N} L \quad \text{for all } z \in \Delta_N, \quad (2)$$

by (iii) and

$$\int_{\gamma_N} (f(Z) + (z - Z)f'(Z))\, dz = 0 \quad (3)$$

by 3.8. Hence, by (1), (2), (3), and 3.10,

$$\left| \int_{\gamma_N} f(z)\, dz \right| \leq \varepsilon(2^{-N}) L \times \text{length}(\gamma_N) = \varepsilon(2^{-N} L)^2.$$

By (iv), $|I| \leq \varepsilon L^2$. Since ε is arbitrary, $I = 0$, as required. $\qquad \square$

4.3 Indefinite-integral theorem (I)

Let f be a continuous complex-valued function on a convex region G such that $\int_\gamma f(z)\,dz = 0$ for any triangle γ in G. Let a be an arbitrary fixed point of G. Then F defined by

$$F(z) = \int_{[a,z]} f(w)\,dw$$

is holomorphic in G, with $F' = f$.

Proof. Fix $z \in G$ and let $D(z;r) \subseteq G$, so that $|h| < r$ implies $z + h \in G$. We compute the limit as $h \to 0$ of $(F(z+h) - F(z))/h$. We shall show that this limit is $f(z)$. For $|h| < r$, the line segments $[a, z]$, $[z, z+h]$, and $[a, z+h]$ all lie in G, since G is convex. By hypothesis, if γ is the triangular contour $[a, z, z+h]$, the integral of f along γ is zero. Hence, by Lemma 3.5,

$$F(z+h) - F(z) = \int_{[a,z+h]} f(w)\,dw - \int_{[a,z]} f(w)\,dw = \int_{[z,z+h]} f(w)\,dw.$$

Also, by parametrization, $\int_{[z,z+h]} 1\,dw = h$. Hence

$$\left| \frac{F(z+h) - F(z)}{h} - f(z) \right| = \frac{1}{|h|} \left| \int_{[z,z+h]} [f(w) - f(z)]\,dw \right|$$

$$\leq \frac{1}{|h|} \times |h| \times \sup_{w \in [z,z+h]} |f(w) - f(z)| \quad \text{(by 3.10)}$$

and this tends to zero as $h \to 0$ (by continuity of f at z). $\qquad\square$

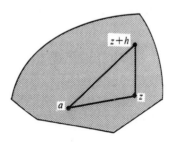

Fig. 4.2

4.4 Antiderivative theorem (I)

Let G be a convex region and $f \in H(G)$. Then there exists $F \in H(G)$ such that $F' = f$.

Proof. Combine Theorems 4.2 and 4.3. □

4.5 Cauchy's theorem for a convex region

Let G be a convex region and let $f \in H(G)$. Then $\int_\gamma f(z)\,dz = 0$ for every closed path γ in G.

Proof. Combine Theorem 4.4 and the Fundamental theorem of calculus, 3.8. □

4.6 Cauchy's theorem (I)

Suppose that f is holomorphic inside and on a contour γ. Then $\int_\gamma f(z)\,dz = 0$.

Proof. Suppose first that γ is a polygon. By triangulating γ (see 3.25), we can write $\int_\gamma f(z)\,dz$ as $\sum_{k=1}^N \int_{\gamma_k} f(z)\,dz$, where each γ_k is a triangle; note that the integrals along the inserted line segments cancel. By Theorem 4.2, the integral of f along each γ_k is zero, so $\int_\gamma f(z)\,dz = 0$.

Now let γ be any contour, and let G be an open set containing $\gamma^* \cup I(\gamma)$ on which f is holomorphic. We shall 'approximate' γ by a polygonal contour. To do this we cover γ^* with overlapping discs $D_k = D(\gamma(t_k); m)$ $(k = 0, \ldots, N; t_0 < t_1 < \ldots < t_N;\;\; \gamma(t_0) = \gamma(t_N))$ which satisfy the conditions (i)–(iv) of the Covering theorem, 3.23. By increasing the number of discs if necessary we may assume that each $\gamma_k := \gamma \restriction [t_k, t_{k+1}]$ is a line segment or a circular arc, and also that the line segments $\tilde\gamma_k := [\gamma(t_k), \gamma(t_{k+1})]$ $(k = 0, \ldots, N-1)$ join to form a polygonal contour $\tilde\gamma$ such that $\tilde\gamma^* \cup I(\tilde\gamma)$ is contained in $\bigcup_{k=0}^N D_k \cup I(\gamma)$, and so in G; see Fig. 4.3. We have $\int_{\tilde\gamma} f(z)\,dz = 0$. Further, for each k, the join of γ_k and $-\tilde\gamma_k$ is a closed path in D_k, which is convex. By Theorem 4.5 and Lemma 3.5,

$$\int_{\gamma_k} f(z)\,dz = \int_{\tilde\gamma_k} f(z)\,dz \quad \text{for each } k.$$

Hence

$$\int_\gamma f(z)\,dz = \sum_{k=0}^{N-1} \int_{\gamma_k} f(z)\,dz = \sum_{k=0}^{N-1} \int_{\tilde\gamma_k} f(z)\,dz = \int_{\tilde\gamma} f(z)\,dz = 0. \quad \square$$

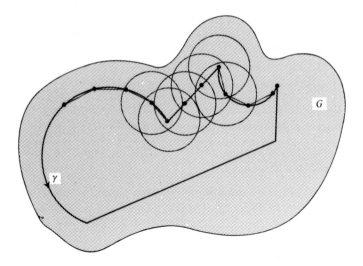

Fig. 4.3

4.7 Orientation

The orientation of a contour has not entered into the theorems proved so far. The next result requires a specific orientation. What is needed is a way of determining orientation consistent with the built-in orientation of a circle $\gamma(a; r)$. We define a contour γ to be *positively oriented* if, as t increases, $\gamma(t)$ moves anticlockwise round any point in $I(\gamma)$. A more formal definition of orientation, in terms of index, is given in 4.20.

4.8 The Deformation theorem (I)

Suppose that γ is a positively oriented contour, that $\bar{D}(a; r) \subseteq I(\gamma)$, and that f is holomorphic inside and on γ except maybe at a. Then

$$\int_{\gamma} f(z)\, dz = \int_{\gamma(a;r)} f(z)\, dz.$$

(The hypotheses are such as to allow us to apply the theorem to functions, such as $1/(z - a)$, which fail to be holomorphic at a. It will be clear that a more general theorem could be proved. Deformation is treated in a much less *ad hoc* way on Level II.)

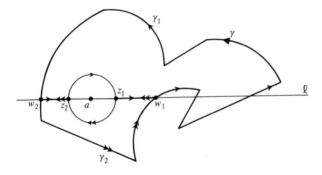

Fig. 4.4

Proof. Take a line ℓ through a which passes through no 'corner point' of γ^* and which is nowhere tangent to γ^*. Let z_1 and z_2 be the points where ℓ meets the circle $|z - a| = r$, and let w_1 and w_2 be the points on $\gamma^* \cap \ell$ such that, for $k = 1$ or 2, z_k lies between a and w_k and $|w_k - a|$ is as small as possible; see Fig. 4.4. Form new contours γ_1 and γ_2 as indicated; each comprises a part of γ, a part of $-\gamma(a; r)$, and line segments joining z_1 to w_1 and z_2 to w_2 with the appropriate orientations. By Theorem 4.6,

$$\int_{\gamma_1} f(z)\, dz = \int_{\gamma_2} f(z)\, dz = 0,$$

and, by Lemma 3.5,

$$\int_{\gamma_1} f(z)\, dz + \int_{\gamma_2} f(z)\, dz = \int_{\gamma} f(z)\, dz - \int_{\gamma(a;r)} f(z)\, dz,$$

since the line segment integrals cancel. □

4.9 Remarks

We have completed the proofs of the major theorems in their Level I forms. Having scaled such a succession of peaks it is perhaps time to pause and take a look at the view.

We began in Chapter 3 by evaluating integrals from scratch by recourse to the parametric definition given in 3.4, and then graduated to the Fundamental theorem of calculus. To use the latter we must recognize our integrand as a continuous derivative, and this may lead to laborious manipulation, or not be feasible at all. With Cauchy's theorem at our disposal we see instantly that our hard-won answer in Example 3.7(2) is right. We also now know, for example, $\int_\gamma e^{z^2} dz = 0$ for *any* contour γ. Try obtaining this without Cauchy's theorem!

4.10 Examples

We claim that $I := \int_{\gamma(0;1)} f(z)\, dz = 0$ for each of the functions f below.

(1) Take $f(z) = 1/z^2$. Then $I = 0$ by 3.7(1).

(2) Take $f(z) = \mathrm{cosec}^2 z$. Then $f(z) = (d/dz)\cot z$ in an open set containing $\gamma(0; 1)^*$, so $I = 0$ by the Fundamental theorem of calculus, 3.8.

(3) Take $f(z) = (4 + z^2)^{-1} e^{iz^2}$. In this case $I = 0$ by Cauchy's theorem I, 4.6.

(4) Take $f(z) = (\operatorname{Im} z)^2$. Then, by 3.4,

$$I = \int_0^{2\pi} (\sin 2t) \mathrm{i} e^{it}\, dt = \int_0^{2\pi} -2 \cos t \sin^2 t\, dt + 2\mathrm{i} \int_0^{2\pi} \sin t \cos^2 t\, dt = 0.$$

(5) Take $f(z) = (2z - 1)^{-1} - (2z + 1)^{-1}$. Direct evaluation of I is unappealing, and 3.8 is inapplicable (why?). But, by Deformation theorem (I), 4.8,

$$I = \int_{\gamma(1/2;1/4)} \frac{1}{2z - 1}\, dz - \int_{\gamma(-1/2;1/4)} \frac{1}{2z + 1}\, dz,$$

and this is $\frac{1}{2} \cdot 2\pi\mathrm{i} - \frac{1}{2} \cdot 2\pi\mathrm{i}$, by 3.7(1).

Note that Cauchy's theorem applies only in case (3). In (1), (2), and (5) the function f fails to be holomorphic at one or more (isolated) points inside $\gamma(0; 1)$. In (4), f is not differentiable anywhere (see Exercise 2.2).

Cauchy's theorem, Level II

4.11 The Deformation theorem (II)

Suppose that f is holomorphic in an open set G and that γ and $\tilde{\gamma}$ are homotopic closed paths in G. Then

$$\int_\gamma f(z)\, dz = \int_{\tilde{\gamma}} f(z)\, dz.$$

Proof. We may assume that $\tilde{\gamma}$ is obtained from γ by an elementary deformation. We adopt the notation of 3.19. For each $k = 0, \ldots, N-1$, the join Γ_k of γ_k, $[\gamma(t_{k+1}), \tilde{\gamma}(s_{k+1})]$, $-\tilde{\gamma}_k$, and $[\tilde{\gamma}(s_k), \gamma(t_k)]$ is a closed path in the convex region G_k, so by

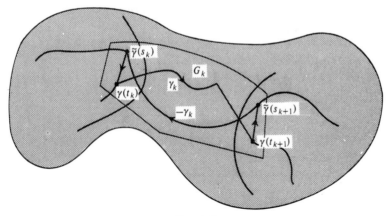

Fig. 4.5

Theorem 4.5, $\int_{\Gamma_k} f(z)\,dz = 0$. Then

$$\int_\gamma f(z)\,dz - \int_{\tilde\gamma} f(z)\,dz = \sum_{k=0}^{N-1}\left(\int_{\gamma_k} f(z)\,dz - \int_{\tilde\gamma_k} f(z)\,dz\right)$$

$$= \sum_{k=0}^{N-1}\int_{\Gamma_k} f(z)\,dz,$$

since the integrals along the line segments cancel. $\qquad\square$

4.12 Cauchy's theorem (II)

Suppose f is holomorphic in a simply connected region G. Then $\int_\gamma f(z)\,dz = 0$ for every closed path γ in G.

Proof. The simple connectedness of G means that γ is homotopic to a null path $\tilde\gamma$ (see 3.20). By the Deformation theorem (II),

$$\int_\gamma f(z)\,dz = \int_{\tilde\gamma} f(z)\,dz,$$

and this is clearly zero. $\qquad\square$

4.13 Postscript to Example 3.21 (2)

The function $(z-a)^{-1}$ is holomorphic in the annulus $A := \{z : R < |z-a| < S\}$ $(0 \le R < S \le \infty)$. The circle $\gamma = \gamma(a;r)$, where $R < r < S$, lies in A and $\int_\gamma (z-a)^{-1}\,dz = 2\pi i \ne 0$. Hence Theorem 4.12 implies that no open annulus is simply connected.

4.14 The Antiderivative theorem (II)

Let G be a simply connected region and let $f \in H(G)$. Then there exists $F \in H(G)$ such that $F' = f$.

Proof. We cannot, as in the proof of Theorem 4.3, define $F(z) = \int_{[a,z]} f(w) \, dw$, since there may not exist a universal point $a \in G$ such that $[a, z] \subseteq G$ for every $z \in G$. The remedy is to substitute for $[a, z]$ some polygonal path $\gamma(z)$ in G joining a fixed point a to z; this is possible by Theorem 3.15. Then, if $D(z; r) \subseteq G$ and $|h| < r$,

$$F(z+h) - F(z) = \int_{\gamma(z+h)} f(w) \, dw - \int_{\gamma(z)} f(w) \, dw$$

$$= \int_{[z, z+h]} f(w) \, dw,$$

by Lemma 3.5 and Theorem 4.12 (Cauchy's theorem (II)). The proof is completed in the same way as that of Theorem 4.4. □

Logarithms, argument, and index

We showed in 2.19 that for each $z \neq 0$ we can find infinitely many solutions to the equation $e^w = z$, differing by integer multiples of $2\pi i$. We now discuss when we can find a holomorphic function f such that $e^{f(z)} = z$ for all z lying in some subset of $\mathbb{C} \backslash \{0\}$.

4.15 Theorem

Suppose G is an open disc [a simply connected region] not containing 0. Then there exists a function $f = \log_G \in H(G)$ such that $e^{f(z)} = z$ for all $z \in G$ and

$$f(z) - f(a) = \int_\gamma \frac{1}{w} \, dw \quad \text{for all } a \text{ and } z \text{ in } G,$$

where γ is any path in G with endpoints a and z. The function f is uniquely determined up to the addition of an integer multiple of $2\pi i$. For each $z \in G$,

$$\log_G z = \log |z| + i\theta(z),$$

where $\theta(z) \in [\arg z]$ and $z \mapsto \theta(z)$ is a continuous function in G.

Proof. By the Antiderivative theorem, 4.4 [4.14], there exists $f \in H(G)$ such that $f'(z) = 1/z$ for all $z \in G$. The derivative of $z e^{-f(z)}$ is, by the chain rule, $e^{-f(z)} - zf'(z)e^{-f(z)}$, and this is zero. Hence $z = Ce^{f(z)}$, where C is a non-zero constant, by 3.18. By adding a suitable constant to f we may assume $C = 1$. The integral formula for f comes from the proof of the Antiderivative theorem. Suppose that for all $z \in G$, $e^{f(z)} = e^{g(z)}$ where f and g belong to $H(G)$. Then $f - g$ has zero derivative and so, by 3.18 again, is a constant K. Since $e^K = 1$, necessarily $K = 2n\pi i$ for some integer n.

The last part comes from 2.19 and the fact that the imaginary part of a holomorphic function must be continuous. □

Theorem 4.15 shows that we can find a continuous argument function in any open disc [or more generally any simply connected region] not containing 0. We shall shortly discuss the variation of argument along an arbitrary closed path γ with $0 \notin \gamma^*$.

4.16 Index

Let γ be a closed path and suppose $w \notin \gamma^*$. Define the *index* (or *winding number*), $n(\gamma, w)$, of γ about w by

$$n(\gamma, w) := \frac{1}{2\pi i} \int_\gamma \frac{1}{z - w} \, dz.$$

Take $\gamma = \gamma(a; r)$. If $|w - a| > r$, $n(\gamma, w) = 0$ (by Cauchy's theorem (I), 4.6). If $|w - a| \leq r$, then the Deformation theorem (I), 4.8, implies that for ε suitably small, $n(\gamma, w) = n(\gamma(w; \varepsilon), w)$. By 3.7(1), $n(\gamma(w; \varepsilon), w) = 1$. Hence $n(\gamma, w) = 1$ if $|w - a| < r$. Also $n(-\gamma, w) = -n(\gamma, w)$ (by 3.5(1)). Thus, at least when γ is a circular contour, the index $n(\gamma, w)$ measures the number of times γ winds round w, taking direction into account. The next theorem shows that such an interpretation is still valid for a general closed path γ. Since for $w \notin \gamma^*$, $n(\gamma, w) = n(\gamma_w, 0)$, where $\gamma_w(t) = \gamma(t) - w$, we may without loss of generality take $w = 0$.

4.17 Theorem

Suppose γ is a closed path with parameter interval $[\alpha, \beta]$ and suppose $0 \notin \gamma^*$. Then
(1) $n(\gamma, 0)$ is an integer, where $2\pi i \, n(\gamma, 0) = \int_\gamma z^{-1} \, dz$.
(2) There exists a continuous function $\eta : [\alpha, \beta] \to \mathbb{R}$, unique up to an integer multiple of 2π, such that
 (i) $2\pi n(\gamma, 0) = \eta(\beta) - \eta(\alpha)$;
 (ii) $\eta(t) \in [\arg \gamma(t)]$ for all $t \in [\alpha, \beta]$.

Proof. Let G be an open set containing γ^* with $0 \notin G$. We should like to use Theorem 4.15, but are prevented from doing so directly because G may not be of the right form. We therefore construct points $\alpha = t_0 < t_1 < \ldots < t_N = \beta$ and discs $D_0, D_1, \ldots, D_N$ as in the Covering theorem, 3.23. For $k = 0, 1, \ldots, N$, let g_k be a holomorphic logarithm in D_k, as constructed in Theorem 4.15. For $z \in D_k$,

$$g_k(z) = \log |z| + i\theta_k(z),$$

and for $z \in D_k \cap D_{k-1}$, $\theta_{k-1}(z) - \theta_k(z) = 2\pi n_k$, where $n_k \in \mathbb{Z}$. Let $z_k = \gamma(t_k)$ $(k = 0, \ldots, N)$. Note that $z_0 = \gamma(\alpha) = \gamma(\beta) = z_N$. Then

$$n(\gamma, 0) = \frac{1}{2\pi i} \int_\gamma \frac{1}{z} \, dz$$

$$= \frac{1}{2\pi i} \sum_{k=0}^{N-1} \int_{\gamma \, [t_k, t_{k+1}]} \frac{1}{z} \, dz \quad \text{(by 3.5(2))}$$

$$= \frac{1}{2\pi i} \sum_{k=0}^{N-1} [g_k(z_{k+1}) - g_k(z_k)] \quad \text{(by Theorem 4.15)}$$

$$= \frac{1}{2\pi} \sum_{k=0}^{N-1} [\theta_k(z_{k+1}) - \theta_k(z_k)] \quad \text{(since the real parts cancel)}$$

$$= \frac{1}{2\pi} \sum_{k=1}^{N} [\theta_{k-1}(z_k) - \theta_k(z_k)] \quad \text{(since } z_0 = z_N\text{)}$$

$$= n_1 + n_2 + \ldots + n_N,$$

which is an integer.

For $k = 0, \ldots, N-1$, define η_k on $[t_k, t_{k+1}]$ by $\eta_k := \theta_k \circ \gamma$; as the composite of continuous functions, η_k is continuous. We patch together the functions η_k to form the function η required in (2), adjusting constants to ensure that η is continuous at the points t_k. The recipe is

$$\eta(t) = \begin{cases} \eta_0(t) & \text{if } t \in [t_0, t_1] \\ \eta_k(t) + \sum_{r=1}^{k} (\eta_{r-1}(t_r) - \eta_r(t_r)) & \text{if } t \in [t_k, t_{k+1}] \quad (1 \leq k \leq N). \end{cases}$$

Finally,

$$2\pi n(\gamma, 0) = \sum_{k=0}^{N-1} \{\theta_k[\gamma(t_{k+1})] - \theta_k[\gamma(t_k)]\} \quad \text{(from above)}$$

$$= \sum_{k=0}^{N-1} [\eta_k(t_{k+1}) - \eta_k(t_k)] \quad \text{(by definition of } \eta_k\text{)}$$

$$= \sum_{k=0}^{N-1} [\eta(t_{k+1}) - \eta(t_k)] \quad \text{(by definition of } \eta\text{)}$$

$$= \eta(\beta) - \eta(\alpha). \qquad \square$$

4.18 Remarks

We call the function η in Theorem 4.17 a *continuous selection of argument along* γ. Note that η is required to vary continuously with t, rather than with $z = \gamma(t)$ (cf. 4.15); when $n(\gamma, 0) \neq 0$, we cannot find a continuous argument function (of z) on γ^* since no choice from $[\arg(z)]$ at $z = \gamma(\alpha) = \gamma(\beta)$ is compatible with continuity. We use this in our discussion of multifunctions in Chapter 6.

Fig. 4.6

4.19 Examples

(1) Let $\gamma(t) = e^{it}$ $(t \in [0, 2\pi])$. For each fixed integer n, $\eta(t) := t + 2\pi n$ gives a continuous selection of argument along γ. Once $\eta(0)$ has been decided the other values of $\eta(t)$ are dictated by the continuity restriction. In particular we must have $\eta(2\pi) = \eta(0) + 2\pi$.

(2) Let $\gamma(t) = 2 + e^{it}$ $(t \in [0, 2\pi])$. In this case a possible choice for η is given by

$$\eta(t) = \tan^{-1}\left(\frac{\sin t}{2 + \cos t}\right),$$

where we take the principal value of $\tan^{-1}$ (having values in $(-\tfrac{1}{2}\pi, \tfrac{1}{2}\pi)$). As t increases from 0 to 2π, $\eta(t)$ increases from 0 to $\pi/6$, decreases from $\pi/6$ to $-\pi/6$, and then increases again to its original value 0.

4.20 Orientation reviewed

Let γ be a closed path with parameter interval $[\alpha, \beta]$. If $n(\gamma, w) = +1$, then as t increases from α to β, $\gamma(t)$ winds once anticlockwise round w, and conversely. This suggests we should replace our interim definition in 4.7 by the following: a contour γ is *positively oriented* if $n(\gamma, w) = +1$ for any $w \in I(\gamma)$.

Cauchy's theorem revisited

Neither Cauchy's theorem (I) nor Cauchy's theorem (II) fully reveals what makes the Cauchy theorems work. To clarify matters we put forward, without proof, a third, and topologically quite sophisticated, Cauchy theorem. An elegant and relatively elementary proof of this result can be found in [7].

4.21 Cauchy's theorem (III)

Suppose G is a region and that $f \in H(G)$. For any closed path γ in G such that $n(\gamma, w) = 0$ for all $w \notin G$, $\int_\gamma f(z)\, dz = 0$.

What is crucial here is the interaction between γ and G via the index. Versions of Cauchy's theorem which do not mention index incorporate geometric restrictions on γ or topological restrictions on G which force the index condition to hold. Intuitively the condition says that γ does not wind round points outside G. Our comments in 4.16 show that Cauchy's theorem (III) is a natural generalization of Cauchy's theorem (I) (for contours). The connection with Cauchy's theorem (II) is more subtle, and proper appreciation of it demands an understanding of algebraic topology (specifically of the relation between homotopy and homology). Some sense of perspective is conveyed by another deep theorem.

4.22 Theorem

Suppose G is a region. Then the following are equivalent:
(1) G is simply connected;
(2) $n(\gamma, w) = 0$ for all $w \notin G$;
(3) $\int_\gamma f(z)\,dz = 0$ for all closed paths γ in G and all $f \in H(G)$;
(4) each $f \in H(G)$ has an antiderivative (that is, $f = F'$ for some $F \in H(G)$);
(5) if $f \in H(G)$, $f : G \to \mathbb{C}\backslash\{0\}$, then f has a holomorphic logarithm (that is, $e^g = f$ for some $g \in H(G)$).

The assertion (1) $\Rightarrow$ (2) is a consequence of the (non-trivial) result that if γ is homotopic to a null path and $w \notin G$, then $n(\gamma, w) = 0$. Conversely (2) $\Rightarrow$ (1) follows, by Cauchy's theorem (II), from the fact that $f(z) = 1/(z - w)$ is holomorphic in G and yet

$$\int_\gamma f(z)\,dz = 2\pi i\, n(\gamma, w) \neq 0.$$

The implication (3) $\Rightarrow$ (4) has already been established, and (4) $\Rightarrow$ (5) is an extension of Theorem 4.15. Completing the circle (by proving (5) $\Rightarrow$ (2)) is much harder, and well beyond the scope of this book. We refer the interested reader to Rudin [7]. We also recommend Beardon [3] to anyone wishing to gain a deeper understanding of index and argument.

Exercises

1. Each of the following integrals is zero:

$$\int_\gamma \frac{1}{z - 2}\,dz \quad \text{(where } \gamma \text{ is a contour with } \gamma^* \subseteq D(0; 1)),$$

$$\int_{\gamma(i;3)} \frac{1}{(z - 2)^3}\,dz, \qquad \int_{\gamma(1;2)} \frac{\sin z}{z}\,dz,$$

$$\int_{\gamma(0;1)} z\,|z|^4\,dz, \qquad \int_{\gamma(1;1)} (1 + e^z)^{-1}\,dz.$$

Give a reason (or reasons) in each case.

2. Evaluate $\int_\gamma (1 + z^2)^{-1} dz$ when γ is (i) $\gamma(1; 1)$, (ii) $\gamma(i; 1)$, (iii) $\gamma(-i; 1)$, (iv) $\gamma(0; 2)$, (v) $\gamma(3i; \pi)$. (Elaborate calculations are unnecessary.)

3. Suppose that f is holomorphic inside and on a regular polygon γ. Using only 4.2 (and any results from Chapter 3 that you need), prove that $\int_\gamma f(z) dz = 0$.

4. Let γ be a polygonal path with initial point 0 and final point 1. What are all the possible values of (i) $\int_\gamma z^3 dz$, (ii) $\int_\gamma (1+z^2)^{-1} dz$? (In (ii), assume $\pm i \notin \gamma^*$.)

5. Define a path γ whose image γ^* is the ellipse

$$\frac{x^2}{a^2} + \frac{y^2}{b^2} = 1,$$

traced anticlockwise. By showing that $\int_\gamma z^{-1} dz = \int_{\bar\gamma} z^{-1} dz$ for a suitable circle $\bar\gamma$, prove that

$$\int_0^{2\pi} \frac{1}{a^2 \cos^2 t + b^2 \sin^2 t} dt = \frac{2\pi}{ab} \qquad (a>0, b>0).$$

6. In $\mathbb{C}_\pi := \{z = |z|e^{i\theta} \neq 0 : -\pi < \theta < \pi\}$, let

$$f(z) := \int_{\gamma(z)} \frac{1}{w} dw,$$

where the path $\gamma(z)$ is the join of $\gamma_1 := [1, |z|]$ and γ_2 defined by

$$\gamma_2(t) = \begin{cases} |z|e^{it} & (t \in [0, \theta]) & \text{if Im } z \geq 0, \\ |z|e^{i(\theta-t)} & (t \in [\theta, 0]) & \text{if Im } z < 0. \end{cases}$$

Verify that, for $z = |z|e^{i\theta} \in \mathbb{C}_\pi$, $f(z) = \log|z| + i\theta$.

7. Suppose f is holomorphic in an open set G and such that $e^g = f$ for some $g \in H(G)$. Prove that, for any given integer $n \geq 2$, there exists $h \in H(G)$ such that $h^n = f$.

8. Prove that it is not possible to find a continuous function θ of z on $\mathbb{C}\backslash\{0\}$ such that $\theta(z) \in [\arg z]$ for all $z \neq 0$. (Hint: apply 1.19 to the function k defined by

$$k(t) = \frac{1}{2\pi}[\theta(e^{it}) + \theta(e^{-it})].$$

5 Consequences of Cauchy's theorem

Armed with Cauchy's theorem we can now prove a host of striking theorems about holomorphic functions. We prove, *inter alia*, that a holomorphic function is automatically infinitely differentiable and locally representable by power series. The reader may also be surprised to discover that:

a bounded function holomorphic in $\mathbb{C}$ is constant (Liouville's theorem, 5.2);

if f is holomorphic in a region G and is zero in an open subdisc of G, then f is identically zero (a corollary of the Identity theorem, given in 5.15);

the modulus of a non-constant holomorphic function cannot attain a maximum at any point of a region (the Maximum-modulus theorem, 5.20).

All this is in sharp contrast to the behaviour of real-valued functions on $\mathbb{R}$, and welcome contrast, since the statements of the theorems are not hedged around with unmemorable technical restrictions. We give a flow chart for the chapter to help the reader follow our route and to highlight the way in which all the results stem from Cauchy's theorem.

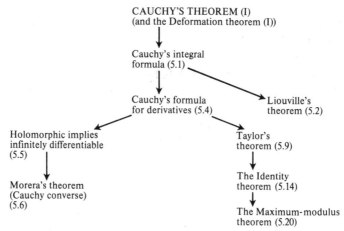

CAUCHY'S THEOREM (I)
(and the Deformation theorem (I))

Cauchy's integral
formula (5.1)

Cauchy's formula
for derivatives (5.4)   Liouville's
                           theorem (5.2)

Holomorphic implies
infinitely differentiable
(5.5)   Taylor's
                                           theorem (5.9)

Morera's theorem
(Cauchy converse)
(5.6)   The Identity
                                           theorem (5.14)

                                           The Maximum-modulus
                                           theorem (5.20)

Table 5.1

Cauchy's formulae

5.1 Cauchy's integral formula

Let f be holomorphic inside and on a positively oriented contour γ. Then, if a is inside γ,

$$f(a) = \frac{1}{2\pi i} \int_\gamma \frac{f(w)}{w - a} \, dw.$$

Proof. There exists r such that $D(a; r) \subseteq I(\gamma)$. For any $\varepsilon < r$,

$$\int_\gamma \frac{f(w)}{w - a} \, dw = \int_{\gamma(a;\varepsilon)} \frac{f(w)}{w - a} \, dw,$$

by the Deformation theorem (I), 4.8. Hence (using Example 3.7(1)),

$$\left| \frac{1}{2\pi i} \int_\gamma \frac{f(w)}{w - a} \, dw - f(a) \right| = \left| \frac{1}{2\pi i} \int_{\gamma(a;\varepsilon)} \frac{f(w) - f(a)}{w - a} \, dw \right|$$

$$= \left| \frac{1}{2\pi i} \int_0^{2\pi} \frac{f(a + \varepsilon e^{i\theta}) - f(a)}{\varepsilon e^{i\theta}} i\varepsilon e^{i\theta} \, d\theta \right|$$

$$\leqslant \frac{1}{2\pi} 2\pi \times \sup_{\theta \in [0, 2\pi]} |f(a + \varepsilon e^{i\theta}) - f(a)|$$
$$\text{(by 3.10)},$$

and this tends to zero as $\varepsilon \to 0$, since f is continuous at a. The expression on the left is independent of ε and so must be zero. $\quad\square$

5.2 Liouville's theorem

Let f be holomorphic and bounded in the complex plane $\mathbb{C}$. Then f is constant.

Proof. Suppose $|f(w)| \leqslant M$ for all $w \in \mathbb{C}$. Fix a and b in $\mathbb{C}$. Take $R \geqslant 2 \max\{|a|, |b|\}$, so that $|w - a| \geqslant \frac{1}{2}R$ and $|w - b| \geqslant \frac{1}{2}R$ whenever $|w| = R$ (by 1.4(3)). By Cauchy's integral formula, applied with $\gamma = \gamma(0; R)$,

$$f(a) - f(b) = \frac{1}{2\pi i} \int_\gamma f(w) \left(\frac{1}{w - a} - \frac{1}{w - b} \right) dw,$$

so

$$|f(a) - f(b)| \leqslant \frac{1}{2\pi} 2\pi R M \frac{|a - b|}{(\frac{1}{2}R)^2} \qquad \text{(by 3.10)}.$$

The right-hand side can be made arbitrarily small by taking R sufficiently large. Hence, for any a and b in $\mathbb{C}$, $f(a) = f(b)$. □

We observed in passing in 2.16 that $\cos z$ is not bounded in $\mathbb{C}$. Liouville's theorem shows that this behaviour is typical of non-constant functions which are holomorphic everywhere.

Liouville's theorem yields an unexpected bonus: an easy proof of the famous result commonly known as the Fundamental theorem of algebra.

5.3 The Fundamental theorem of algebra

Let $p(z)$ be a non-constant polynomial with complex coefficients. Then there exists $\zeta \in \mathbb{C}$ such that $p(\zeta) = 0$.

Proof. Suppose for a contradiction that $p(z) \neq 0$ for every z. Since $|p(z)| \to \infty$ as $|z| \to \infty$, there exists R such that $1/|p(z)| < 1$ for $|z| > R$. On the compact set $\bar{D}(0; R)$, $1/p(z)$ is continuous and hence bounded (by 1.18). Hence $1/p(z)$ is bounded on $\mathbb{C}$. It is also holomorphic (see 2.7), hence must be constant, by Liouville's theorem. □

The inductive proof given of the next result is less formidable than it may look. Its ingredients are simply Cauchy's integral formula, the definition of derivative, and the Fundamental theorem of calculus. A somewhat simpler proof can be given of the formula for the first derivative (see Exercise 5.8).

5.4 Cauchy's formula for derivatives

Let f be holomorphic inside and on a positively oriented contour γ and let a lie inside γ. Then $f^{(n)}(a)$ exists for $n = 1, 2, \ldots$ and

$$f^{(n)}(a) = \frac{n!}{2\pi i} \int_\gamma \frac{f(w)}{(w-a)^{n+1}}\, dw. \qquad (*)$$

Proof. For $n = 0$, $(*)$ is Cauchy's integral formula. By induction, it will now be sufficient to assume $(*)$ holds for $n = k$, and prove it holds for $n = k+1$. By the Deformation theorem (I), 4.8, we may assume that γ is $\gamma(a; 2r)$, for some constant r. Take $|h| < r$. By $(*)$, for $n = k$,

$$f^{(k)}(a+h) - f^{(k)}(a) = \frac{k!}{2\pi i} \int_\gamma f(w)\left(\frac{1}{(w-a-h)^{k+1}} - \frac{1}{(w-a)^{k+1}}\right) dw$$

$$= \frac{(k+1)!}{2\pi i} \int_\gamma f(w)\left(\int_{[a,\,a+h]} (w-\zeta)^{-k-2}\, d\zeta\right) dw,$$

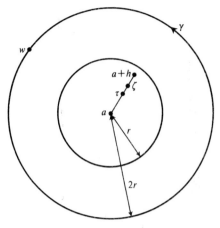

Fig. 5.1

by 3.8. We shall show that $F(h) \to 0$ as $h \to 0$, where

$$F(h) = \frac{f^{(k)}(a+h) - f^{(k)}(a)}{h} - \frac{(k+1)!}{2\pi i} \int_\gamma \frac{f(w)}{(w-a)^{k+2}} \, dw$$

$$= \frac{(k+1)!}{2\pi i h} \int_\gamma f(w) \left(\int_{[a, a+h]} [(w-\zeta)^{-k-2} - (w-a)^{-k-2}] \, d\zeta \right) dw$$

$$= \frac{(k+2)!}{2\pi i h} \int_\gamma f(w) \left[\int_{[a, a+h]} \left(\int_{[a, \zeta]} (w-\tau)^{-k-3} \, d\tau \right) d\zeta \right] dw.$$

Since f is holomorphic and so continuous, it is bounded, by M say, on the compact set γ^* (by 1.18). For $\tau \in [a, \zeta]$ and $\zeta \in [a, a+h]$, we have $|w - \tau| \geq r$ for all $w \in \gamma^*$. Also $|\zeta - a| \leq |h|$ (see Fig. 5.1). Hence, by 3.10, $|F(h)| \leq \dfrac{(k+2)!}{2\pi |h|} \times \dfrac{M|h|^2}{r^{k+3}} \times 4\pi r$, so $F(h) \to 0$ as $h \to 0$. $\qquad \square$

5.5 Corollary

Suppose f is holomorphic in an open set G. Then f has derivatives of all orders in G.

Proof. In 5.4, take $a \in G$ and γ a circle in G enclosing a. $\qquad \square$

Note that the special case of 5.4 used to prove the corollary does not require the Deformation theorem. Corollary 5.5 includes the statement that the derivative of a holomorphic function is holomorphic. From this follows a partial converse to Cauchy's theorem.

5.6 Morera's theorem

Suppose that f is continuous in an open set G and that $\int_\gamma f(z)\,dz = 0$ for all triangles γ in G. Then $f \in H(G)$.

Proof. Let $a \in G$ and choose r such that $D(a; r) \subseteq G$. Since $D(a; r)$ is a convex region, we can, by Theorem 4.3, find $F \in H(D(a; r))$ such that $F' = f$. By Corollary 5.5, $f \in H(D(a; r))$. Since a is arbitrary, $f \in H(G)$. $\qquad\square$

We conclude this section with some examples, the first of which illustrates how our repertoire of techniques for evaluating integrals has been enlarged by the Cauchy formulae. We now know that, given a contour γ, a point $a \notin \gamma^*$, a function f holomorphic inside and on γ and $n = 0, 1, \ldots,$

$$\int_\gamma \frac{f(z)}{(z-a)^{n+1}}\,dz = \begin{cases} 2\pi i f(a) & \text{if } n=0 \text{ and } a \in I(\gamma) \quad \text{(by 5.2)}, \\[2mm] \dfrac{2\pi i}{n!} f^{(n)}(a) & \text{if } n \geqslant 1 \text{ and } a \in I(\gamma) \quad \text{(by 5.4)}, \\[2mm] 0 & \text{if } a \in O(\gamma) \quad \text{(by 4.6)}. \end{cases}$$

5.7 Examples

(1) By 5.1,

$$\int_{\gamma(i; 1)} \frac{z^2}{z^2+1}\,dz = \left[2\pi i\,\frac{z^2}{z+i}\right]_{z=i} = -\pi.$$

(2) By 5.4,

$$\int_{\gamma(0; 1)} e^z z^{-3}\,dz = \left[\frac{2\pi i}{2!}\frac{d^2}{dz^2}(e^z)\right]_{z=0} = \pi i.$$

(3) We cannot evaluate

$$\int_{\gamma(0; 1)} \frac{\operatorname{Re} z}{z - \tfrac{1}{2}}\,dz$$

directly by Cauchy's integral formula, 5.1, since $\operatorname{Re} z$ is not holomorphic (because it fails to satisfy the Cauchy–Riemann equations, 2.4). However, when $|z| = 1$, $z = e^{it}$ and so $\operatorname{Re} z = \cos t = \tfrac{1}{2}(e^{it} + e^{-it}) = \tfrac{1}{2}(z + z^{-1})$. The required integral is therefore, by 5.1,

$$\int_{\gamma(0; 1)} \frac{z^2+1}{2z(z-\tfrac{1}{2})}\,dz = \int_{\gamma(0; 1)} \left(\frac{1}{2}\frac{1}{z} - \frac{1}{z} + \frac{5}{2(2z-1)}\right)\,dz$$
$$= 0 - 2\pi i + 5\pi i/2 = \tfrac{1}{2}\pi i.$$

5.8 Example (the Poisson integral formula)

To prove that, if f is holomorphic inside and on $\gamma(0; 1)$, then

$$f(re^{i\theta}) = \frac{1}{2\pi} \int_0^{2\pi} \frac{1-r^2}{1-2r\cos(\theta-t)+r^2} f(e^{it})\, dt \qquad (re^{i\theta} \in D(0; 1)).$$

Solution. We fix $z = re^{i\theta}$ and apply 5.1 to the product $f\phi$, where $\phi(w) = (1-r^2)/(1-w\bar{z})$. Then

$$f(z) = f(z)\phi(z) = \frac{(1-r^2)}{2\pi i} \int_{\gamma(0;\, 1)} \frac{f(w)}{(w-z)(1-w\bar{z})}\, dw$$

$$= \frac{(1-r^2)}{2\pi i} \int_0^{2\pi} \frac{f(e^{it})ie^{it}}{(e^{it}-re^{i\theta})(1-re^{i(t-\theta)})}\, dt,$$

which simplifies to give the integral required. $\qquad\square$

Power series representation

Theorem 2.12 shows that a convergent complex power series defines a holomorphic function. Theorem 5.9 provides the converse. The two results combine to demonstrate that a function is holomorphic in an open set G if and only if it is *analytic*, that is, locally representable by power series. It is thanks to Corollary 5.5 that Theorem 5.9 is stronger and more satisfactory than most forms of Taylor's theorem for a function of a real variable.

5.9 Taylor's theorem

Let $f \in H(D(a; R))$. Then there exist unique constants c_n such that

$$f(z) = \sum_{n=0}^{\infty} c_n(z-a)^n \qquad (z \in D(a; R)).$$

The constant c_n is given by

$$c_n = \frac{1}{2\pi i} \int_\gamma \frac{f(w)}{(w-a)^{n+1}}\, dw = f^{(n)}(a)/n!,$$

where γ is a circle $\gamma(a; r)$ $(0<r<R)$ [or any contour in $D(a; R)$ homotopic to such a circle].

Proof. Fix $z \in D(a; R)$ and choose r such that $|z-a|<r<R$. Take $\gamma = \gamma(a; r)$. Then

$$f(z) = \frac{1}{2\pi i} \int_\gamma \frac{f(w)}{w-z}\, dw \qquad \text{(by 5.1).}$$

Since $|z-a|<|w-a|$ for all $w \in \gamma^*$, the right-hand side of the equation

$$\frac{1}{w-z} = \frac{1}{w-a} \frac{1}{1-[(z-a)/(w-a)]}$$

can be expanded binomially (recall 2.8), to give

$$f(z) = \frac{1}{2\pi i} \int_\gamma \sum_{n=0}^\infty \frac{(z-a)^n}{(w-a)^{n+1}} f(w) \, dw.$$

On the compact set γ^*, the continuous function f is bounded, so, for some constant M, we have

$$\left| \frac{(z-a)^n}{(w-a)^{n+1}} f(w) \right| \leq \frac{M}{r} \left(\frac{|z-a|}{r} \right)^n =: M_n.$$

$\sum M_n$ converges, since $|z-a| < r$. Hence, by Theorem 3.13 [uniform convergence], summation and integration may be interchanged, to give

$$f(z) = \sum_{n=0}^\infty \left(\frac{1}{2\pi i} \int_\gamma \frac{f(w)}{(w-a)^{n+1}} \, dw \right) (z-a)^n.$$

The assertions of the theorem now follow from Cauchy's formula for derivatives, 5.4, and Corollary 2.13, which implies that the coefficients in the expansion are unique. ☐

5.10 Example

The function f is holomorphic in $\mathbb{C}$. Prove that, if there exist positive constants M and K and a positive integer k such that $|f(z)| \leq M |z|^k$ for $|z| \geq K$, then f is a polynomial of degree at most k.

Solution. By Taylor's theorem, f has a power series expansion $f(z) = \sum_{n=0}^\infty c_n z^n$, valid in every disc with centre 0, where

$$c_n = \frac{1}{2\pi i} \int_{\gamma(0; R)} f(z) z^{-n-1} \, dz.$$

Hence, taking $R \geq K$ and using 3.10 and the given growth condition on f,

$$|c_n| \leq \frac{1}{2\pi} \sup\{|f(z) z^{-n-1}| : |z| = R\} \times \text{length}(\gamma(0; R))$$

$$\leq \frac{1}{2\pi} M R^{k-n-1} \times 2\pi R.$$

Since R can be chosen arbitrarily large, we must have $c_n = 0$ for $n > k$, whence f is a polynomial of degree not greater than k. ☐

The benefits of the complex form of Taylor's theorem are further demonstrated by the following applications.

5.11 Multiplication of power series

Formally

$$\sum_{n=0}^{\infty} a_n z^n \sum_{n=0}^{\infty} b_n z^n = \sum_{n=0}^{\infty} c_n z^n,$$

where $c_n = \sum_{r=0}^{n} a_r b_{n-r}$, the expression for c_n being obtained by noting that terms in z^n arise as products $a_r z^r \times b_s z^s$ where $r+s = n$. The formula is true whenever the series being multiplied converge absolutely. Most proofs of this result are highly technical. Taylor's theorem allows us to supply a neat proof.

Proposition

Suppose $f(z) = \sum_{n=0}^{\infty} a_n z^n$ and $g(z) = \sum_{n=0}^{\infty} b_n z^n$ are complex power series with radii of convergence R_1 and R_2 respectively. Then $h(z) = \sum_{n=0}^{\infty} c_n z^n$, where $c_n = \sum_{r=0}^{n} a_r b_{n-r}$, has radius of convergence at least $R := \min\{R_1, R_2\}$, and $h(z) = f(z)g(z)$ for $|z| < R$.

Proof. In $D(0; R)$, f and g are both holomorphic, and we have $a_n = f^{(n)}(0)/n!$ and $b_n = g^{(n)}(0)/n!$ (by 2.12 and 2.13). The product fg is also holomorphic in $D(0; R)$ and is represented there by a Taylor series

$$f(z)g(z) = \sum_{n=0}^{\infty} c_n z^n,$$

where

$$c_n = \frac{(fg)^{(n)}(0)}{n!} = \sum_{r=0}^{n} \frac{1}{n!} \frac{n!}{r!(n-r)!} f^{(r)}(0) g^{(n-r)}(0) = \sum_{r=0}^{n} a_r b_{n-r},$$

by Leibniz's formula for the nth derivative of a product.   □

5.12 Examples

(1) The exponential series has infinite radius of convergence. The proposition above allows us to multiply the series for e^{az} and e^{bz}. Doing so and putting $z = 1$ gives $e^{a+b} = e^a e^b$ for all a and b in $\mathbb{C}$ (cf. 2.15).

(2) The nth Hermite function H_n is defined by

$$H_n(t) := (-1)^n e^{\frac{1}{2}t^2} \left(\frac{d}{dt}\right)^n (e^{-t^2}) \qquad (n = 0, 1, 2, \ldots).$$

Prove that

$$\sum_{n=0}^{\infty} H_n(t) \frac{x^n}{n!} = e^{-\frac{1}{2}t^2 + 2xt - x^2} \quad (x, t \in \mathbb{R}).$$

Solution. Note that $e^{-\frac{1}{2}t^2 + 2xt - x^2} = e^{\frac{1}{2}t^2} e^{-(x-t)^2}$. By the chain rule, e^{-z^2} is holomorphic in $\mathbb{C}$. It has, for any a, a Taylor expansion

$$e^{-z^2} = \sum_{n=0}^{\infty} \left[\left(\frac{d}{dz}\right)^n (e^{-z^2}) \right]_{z=a} \frac{(z-a)^n}{n!}.$$

If we put $z = x - t$ and $a = -t$ the required formula drops out. $\square$

Note The Hermite functions are of importance in mathematical physics and elsewhere. Differentiation of the series $\sum H_n(t) x^n / n!$ with respect to both x and t can be shown to be legitimate. This enables the generating function constructed above to be used to produce, painlessly, assorted recurrence relations and the differential equation $H_n''(t) = (t^2 - 2n - 1) H_n(t)$.

Zeros of holomorphic functions

Knowledge of the values of a holomorphic function f on, for example, some line segment $[a, p]$ $(p \neq a)$ enables the derivatives $f'(z)$, $f''(z)$, ... for $z \in [a, p]$ to be computed in turn (see 2.3). The values $f(a)$, $f'(a)$, $f''(a)$, ... uniquely determine the values of f in any disc $D(a; r)$ in which f is holomorphic (by 5.9). Hence two functions f and g which are holomorphic in a disc $D(a; r)$ and equal on some segment $[a, p]$ must be equal everywhere in the disc. We can improve substantially on this *ad hoc* uniqueness theorem. To do so we consider zeros of holomorphic functions, motivated by the observation that $f(z) = g(z)$ if and only if $(f - g)(z) = 0$. We shall need two topological concepts introduced earlier: that of a limit point (from 1.10) and of a region (from 3.14).

5.13 Definitions

Suppose f is holomorphic in $D(a; r)$ for some r. The point a is said to be a *zero* of f if $f(a) = 0$. The zero a is *isolated* if there exists ε such that $D'(a; \varepsilon)$ contains no zeros of f.

5.14 The Identity theorem

Let G be a region and suppose $f \in H(G)$. If $Z(f)$, the set of zeros of f in G, has a limit point in G, then f is identically zero in G. (An equivalent formulation is that, in a region, the zeros of a holomorphic function are isolated unless the function is identically zero.)

Proof.

Stage 1 Let $a \in G$, $f(a) = 0$. Take r such that $D(a; r) \subseteq G$. By Taylor's theorem, 5.9,

$$f(z) = \sum_{n=0}^{\infty} c_n (z-a)^n \qquad (z \in D(a; r)).$$

There are two possibilities.

(i) All the coefficients $c_n = 0$, in which case $f \equiv 0$ in $D(a; r)$.

(ii) There exists a smallest integer $m > 0$ such that $c_m \neq 0$. The series $\sum_{n=0}^{\infty} c_{n+m} (z-a)^n$ has radius of convergence at least r, and defines a function g which is continuous in $D(a; r)$ (by 2.12 and 2.3(1)). Because $g(a) \neq 0$ and g is continuous at a, $g(z) \neq 0$ for z in some disc $D(a; \varepsilon)$. In the punctured disc $D'(a; \varepsilon)$, $f(z) = (z-a)^m g(z)$ is never zero, so a is not a limit point of $Z(f)$.

We conclude that if a is a zero of f in G, then *either* $f \equiv 0$ in some disc $D(a; r) \subseteq G$ *or* a is not a limit point of zeros. This completes the proof for the special case that G is a disc and provides a necessary stepping stone in the general case.

Stage 2 Our strategy is now to prove that E, the set of limit points of $Z(f)$, is such that

(i) $Z(f) \supseteq E$, and

(ii) E and $G \backslash E$ are open.

Since G is a region, (ii) implies that $E = G$ (whence, by (i), $f \equiv 0$ in G) or that $E = \varnothing$ (whence $Z(f)$ has no limit points in G).

Suppose, for a contradiction, $a \in E \backslash Z(f)$. Then for each n there exists $a_n \in D'(a; 1/n)$ such that $f(a_n) = 0$. By continuity of f, $f(a) = 0$, contrary to hypothesis.

To prove (ii), first let $a \in E$. Then Stage 1 implies $f \equiv 0$ in some $D(a; r)$. Then $D(a; r) \subseteq E$, so E is open. To show $G \backslash E$ is open, take $a \in G \backslash E$. Since a is not a limit point of $Z(f)$, there exists a disc $D(a; r)$ in which f is never zero. By (i), $E \subseteq Z(f)$. Hence $D(a; r) \subseteq G \backslash E$. $\square$

5.15 Corollary

Suppose G is a region, $f \in H(G)$, and f is identically zero in some $D(a;r) \subseteq G$. Then f is identically zero in G.

Note In Theorem 5.14 and Corollary 5.15, the requirement that G be a region is clearly necessary: consider a function taking different constant values on two disjoint open discs.

The promised uniqueness theorem is an immediate consequence of Theorem 5.14.

5.16 The Uniqueness theorem

Suppose G is a region, f and g belong to $H(G)$, and $f(z) = g(z)$ for all $z \in S \subseteq G$, where S has a limit point in G. Then $f \equiv g$ in G.

5.17 Examples

(1) Suppose that f is holomorphic in $\mathbb{C}$ and that $f(1/n) = \sin(1/n)$ $(n = 1, 2, \ldots)$. The set $S = \{1/n : n = 1, 2, \ldots\}$ has a limit point, 0, and $f(z) = \sin z$ on S. Hence, by the Uniqueness theorem, $f(z) = \sin z$ for all $z \in \mathbb{C}$.

(2) Suppose f is holomorphic in $\mathbb{C}\backslash\{0\}$ and $f(z) = \sin(1/z)$ whenever $z = 1/(n\pi)$ $(n = 1, 2, \ldots)$. It does not follow that necessarily $f(z) = \sin(1/z)$ for $z \neq 0$. Indeed, $f \equiv 0$ would fit the given conditions. Here 0 is a limit point of zeros of $f(z) - \sin(1/z)$, but is not in $\mathbb{C}\backslash\{0\}$, the region of holomorphy of this function.

5.18 Preservation of functional identities

The Uniqueness theorem, 5.16, allows us to extend the domain of validity of certain functional identities, a procedure we alluded to in 2.16. The method is best illustrated by examples.

Assuming the identity $\cos^2 z + \sin^2 z = 1$ holds when z is real, 5.16 implies that it holds for all complex z (since 1 and $\cos^2 z + \sin^2 z$ are holomorphic in $\mathbb{C}$, and the real axis has limit points in $\mathbb{C}$ (in fact every point of $\mathbb{R}$ is a limit point).

Our second example concerns the binomial expansion. Given a negative integer n, let

$$f(z) = (1+z)^n \quad \text{and} \quad g(z) = \sum_{k=0}^{\infty} \frac{n(n-1)\ldots(n-k+1)}{k!} z^k.$$

Clearly f is holomorphic except at -1. The series defining g has radius of convergence 1, so $g \in H(D(0; 1))$, by 2.12. It is well known that $f(x) = g(x)$ when x is real and $|x| < 1$. Theorem 5.16 tells us $f(z) = g(z)$ throughout the region $D(0; 1)$. The requirement that n be an integer is not essential, but cannot be removed until we have developed techniques for handling multifunctions.

The Maximum-modulus theorem

We round off this chapter with another major result, the Maximum-modulus theorem. We present it here as an end in itself. In the exercises and in Chapter 10 we shall see just a few of its applications. Many theorems of advanced complex analysis make use of it in their proofs.

5.19   The Local maximum-modulus theorem

Suppose $f \in H(D(a; R))$ and that $|f(z)| \leqslant |f(a)|$ for all $z \in D(a; R)$. Then f is constant.

Proof. Fix r with $0 < r < R$. By Cauchy's integral formula, 5.1,

$$f(a) = \frac{1}{2\pi i} \int_{\gamma(a; r)} \frac{f(z)}{z - a} \, dz$$

$$= \frac{1}{2\pi i} \int_0^{2\pi} \frac{f(a + re^{i\theta}) r i e^{i\theta}}{re^{i\theta}} \, d\theta$$

$$= \frac{1}{2\pi} \int_0^{2\pi} f(a + re^{i\theta}) \, d\theta.$$

From this and from the hypothesis of the theorem we see that

$$|f(a)| \leqslant \frac{1}{2\pi} \int_0^{2\pi} |f(a + re^{i\theta})| \, d\theta \leqslant |f(a)|,$$

whence

$$\int_0^{2\pi} [|f(a)| - |f(a + re^{i\theta})|] \, d\theta = 0.$$

Since the integrand is continuous and non-negative, it must be identically zero if the integral is to vanish, and this is true for every $r < R$. It follows that $|f|$ is constant in $D(a; R)$. By Proposition 2.6, f itself must be constant. $\square$

5.20 The Maximum-modulus theorem

Let G be a bounded region and let f be holomorphic in G and continuous on $\bar{G}$. Then $|f|$ attains its maximum on the boundary of G, that is, on $\partial G := \bar{G} \backslash G$.

Proof. The set $\bar{G}$ is bounded and closed, so on $\bar{G}$ the continuous function $|f|$ is bounded and attains its supremum M at some point of $\bar{G}$ (by 1.18). We assume $|f|$ does not attain M on ∂G. Then $|f(a)| = M$ for some $a \in G$. By 5.19, f is constant in some disc $D(a; R) \subseteq G$. Hence f is constant in G, by the Identity theorem (see Corollary 5.15). By continuity, f is constant on $\bar{G}$, and so attains its supremum on ∂G, contrary to hypothesis. $\qquad\square$

Exercises

1. Evaluate, when $\gamma = \gamma(0; 2)$,

$$\text{(i)} \int_\gamma \frac{z^3+5}{z-i}\, dz, \quad \text{(ii)} \int_\gamma \frac{1}{z^2+z+1}\, dz, \quad \text{(iii)} \int_\gamma \frac{\sin z}{z^2+1}\, dz,$$

$$\text{(iv)} \int_\gamma (z-1)^{-3} e^{z^2}\, dz, \quad \text{(v)} \int_\gamma z^{-n} \cos z\, dz \; (n = 1, 2, \ldots),$$

$$\text{(vi)} \int_\gamma z^n (1-z)^m\, dz \; (m = 0, 1, 2, \ldots, \; n = 0, \pm1, \pm2, \ldots).$$

2. Suppose f is holomorphic inside and on $\gamma(0; 1)$. Prove that

$$2\pi i f(z) = \int_{\gamma(0;1)} \frac{f(w)}{w-z}\, dw - \int_{\gamma(0;1)} \frac{f(w)}{w - 1/\bar{z}}\, dw \qquad (0 < |z| < 1),$$

and hence give an alternative proof of the Poisson integral formula obtained in 5.8.

3. By considering its complex conjugate, evaluate the integral

$$\frac{1}{2\pi i} \int_{\gamma(0;1)} \frac{\overline{f(z)}}{z-a}\, dz$$

in the cases (i) $|a| < 1$, (ii) $|a| > 1$, where $f \in H(D(0; R))$ $(R > 1)$.

4. Use Liouville's theorem to prove that if f is holomorphic in $\mathbb{C}$ and satisfies $f(z + 2\pi) = f(z)$ and $f(z + 2\pi i) = f(z)$ for all z, then f is constant.

5. Find an expansion $f(z) = \sum_{n=0}^\infty c_n (z-a)^n$ valid in the disc $D(a; r)$ when (i) $f(z) = \sin^2 z$, $a = 0$, (ii) $f(z) = (1+z)^{-1}$, $a = i$, (iii) $f(z) = e^z$, $a = 1$. How big can r be in each case?

6. Suppose $f(z) = \sum_{n=0}^\infty c_n z^n$ for $z \in \mathbb{C}$. Prove that, for all R,

$$\sum_{n=0}^\infty |c_n| R^n \leqslant 2M(2R), \quad \text{where} \quad M(r) := \sup\{|f(z)| : |z| = r\}.$$

7. Let $f \in H(D(0; R))$ and suppose f has Taylor series $\sum_{n=0}^{\infty} c_n z^n$. Use the integral formula for c_n, and the fact that $\int_\gamma f(z) z^{n-1} dz = 0$ for $n = 1, 2, \ldots$ and suitable γ, to show that

$$c_n = \frac{r^{-n}}{\pi} \int_0^{2\pi} \mathrm{Re}[f(re^{i\theta})] e^{-in\theta} d\theta \qquad (n \geqslant 1, 0 < r < R).$$

Deduce that if $\mathrm{Re}\, f \geqslant 0$ in $D(0; R)$ and $f(0) = 1$, then $|c_n| \leqslant 2/R^n$ $(n \geqslant 1)$.

8. Let γ be a path and let $g : \gamma^* \to \mathbb{C}$ be continuous. Define f by

$$f(z) = \int_\gamma \frac{g(w)}{w - z} dw \quad (z \notin \gamma^*).$$

Show that for $z, z + h \notin \gamma^*$,

$$\frac{f(z+h) - f(z)}{h} - \int_\gamma \frac{g(w)}{(w-z)^2} dw = h \int_\gamma \frac{g(w)}{(w - z - h)(w - z)^2} dw$$

and, by estimating the integral on the right-hand side, deduce that f is holomorphic in $\mathbb{C} \setminus \gamma^*$, with

$$f'(z) = \int_\gamma \frac{g(w)}{(w - z)^2} dw.$$

(Hint: Use Exercise 3.12 in the estimation.)

9.   Suppose G is an open set, $f : G \to \mathbb{C}$ is continuous, and f is holomorphic in $G \setminus [a, b]$, where $[a, b]$ is a line segment lying in G. Use Morera's theorem to prove that f is holomorphic in G.

10. Is it possible to construct $f \in H(D(0; 1))$ such that $f(1/n) = z_n$ $(n = 1, 2, \ldots)$ when (i) $z_n = (-1)^n$, (ii) $z_n = n/(n+1)$, (iii) $z_n = 0$ (n even), $z_n = 1/n$ (n odd)?

11. Let $f \in H(D(0; 1))$. Prove that g defined by

$$g(z) = f(z) - \overline{f(-\bar{z})}$$

is holomorphic in $D(0; 1)$. Now suppose that f takes real values on the imaginary axis. Prove that, for $x + iy \in D(0; 1)$,

$$u(x, y) = u(-x, y) \quad \text{and} \quad v(x, y) = -v(-x, y),$$

where u and v denote the real and imaginary parts of f (as in 2.4).

12. Let f be holomorphic in a region G. By considering e^f show that $\mathrm{Re}\, f$ cannot attain a maximum value at any point of G. Can $\mathrm{Re}\, f$ attain a minimum value?

13. Let G be the square region $\{z : |\mathrm{Re}\, z| < 1, |\mathrm{Im}\, z| < 1\}$. Suppose f is continuous in $\bar{G}$, holomorphic in G, and such that $f(z) = 0$ when $\mathrm{Re}\, z = 1$. By considering g defined by $g(z) = f(z)f(iz)f(-z)f(-iz)$, prove that f is identically zero in $\bar{G}$.

14. Let $f \in H(D(0; R))$ and let $M(r) := \sup\{|f(z)| : |z| = r\}$ $(r < R)$. Prove that $M(r) \leqslant M(s)$ for $r < s < R$, with strict inequality if f is non-constant. Prove further that if f is a polynomial of degree n, then $M(r)r^{-n} \geqslant M(s)s^{-n}$ when $0 < r < s < R$.

[The remaining exercises assume familiarity with uniform convergence.

15. Let G be open and suppose $\langle f_k \rangle$ is a sequence of functions such that $f_k \in H(G)$ $(k = 1, 2, \ldots)$ and $f_k \to f$ uniformly on G. Use Morera's theorem to prove that $f \in H(G)$. Use Cauchy's formula for derivatives to prove that

$$f^{(n)}(a) = \lim_{k \to \infty} f_k^{(n)}(a) \quad \text{for each} \quad a \in G, \ n = 1, 2, \ldots .$$

Deduce corresponding results for uniformly convergent series of holomorphic functions.

16. Prove that, for each $\delta > 0$, $\sum_{n=1}^{\infty} n^{-z}$ converges uniformly on $\{z : \operatorname{Re} z > 1 + \delta\}$. Deduce that the series defines a holomorphic function $\zeta(z)$ in $\{z : \operatorname{Re} z > 1\}$. (This is the *Riemann zeta function*, of great importance in number theory.)

17. By proving that the series converges uniformly on any disc $D(a; r)$ containing no integer, prove that $\sum_{n=-\infty}^{\infty} (z - n)^{-2}$ defines a function holomorphic in $\mathbb{C} \setminus \mathbb{Z}$. (In neither of the last two exercises does the given series converge uniformly on the whole of the region of holomorphy.)]

6 Singularities and multifunctions

The time has come to face the fact that many commonplace functions fail to be holomorphic at isolated points (for example $1/z$ at 0) or (like the logarithm) are not proper functions at all, because they are many-valued. Laurent's theorem, 6.1, provides a very satisfactory substitute for Taylor's theorem for functions holomorphic except at isolated singular points. It also leads to a classification of singularities which proves invaluable in Chapters 7–9. The final section of this chapter discusses some important multifunctions, and shows how they can be doctored to bring them within the scope of Cauchy's theorem and its consequences.

Laurent's theorem

If f is holomorphic in a disc $D(a \,; r)$, except at a itself where something nasty happens, we cannot hope for a power series expansion

$$f(z) = \sum_{n=0}^{\infty} c_n(z-a)^n,$$

since the right-hand side of this behaves decently at a, by Theorem 2.12. We aim instead for an expansion

$$f(z) = \sum_{n=-\infty}^{\infty} c_n(z-a)^n,$$

valid for $0 < |z-a| < r$ (that is, for $z \in D'(a \,; r)$). We prove, slightly more generally, that a function holomorphic in an open annulus has an expansion of this type.

6.1 Laurent's theorem

Let $A = \{z : R < |z-a| < S\}$ $(0 \leqslant R < S \leqslant \infty)$ and let $f \in H(A)$. Then

$$f(z) = \sum_{n=-\infty}^{\infty} c_n(z-a)^n \qquad (z \in A),$$

where

$$c_n = \frac{1}{2\pi i} \int_\gamma \frac{f(w)}{(w-a)^{n+1}} \, dw,$$

with $\gamma = \gamma(a; r)$ $(R < r < S)$.

Note By definition, a double-ended series $\sum_{n=-\infty}^{\infty} a_n$ converges (to $s = s_1 + s_2$) if $\sum_{n=0}^{\infty} a_n$ converges (to s_1) and $\sum_{n=1}^{\infty} a_{-n}$ converges (to s_2).

Proof. By changing the origin if necessary, we may assume $a = 0$. Fix $z \in A$ and choose P and Q such that $R < P < |z| < Q < S$. Let $\tilde{\gamma}$ and $\tilde{\tilde{\gamma}}$ be as shown in Fig. 6.1. Then

$$f(z) = \frac{1}{2\pi i} \int_{\tilde{\gamma}} \frac{f(w)}{w - z} \, dw \quad \text{(by 5.1)},$$

and

$$0 = \frac{1}{2\pi i} \int_{\tilde{\tilde{\gamma}}} \frac{f(w)}{w - z} \, dw \quad \text{(by Cauchy's theorem (I))}.$$

Adding these, the integrals along the line segments cancel, so

$$f(z) = \frac{1}{2\pi i} \int_{\gamma(0;\,Q)} \frac{f(w)}{w - z} \, dw - \frac{1}{2\pi i} \int_{\gamma(0;P)} \frac{f(w)}{w - z} \, dw.$$

$$= \frac{1}{2\pi i} \int_{\gamma(0;Q)} \sum_{n=0}^{\infty} \frac{z^n}{w^{n+1}} f(w) \, dw - \frac{1}{2\pi i} \int_{\gamma(0;P)} \sum_{m=0}^{\infty} -\frac{w^m}{z^{m+1}} f(w) \, dw,$$

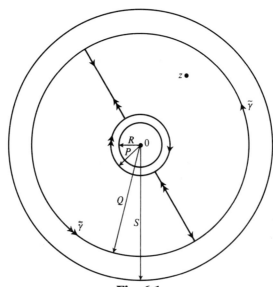

Fig. 6.1

using the appropriate binomial expansions; we have $|z/w|<1$ for $w \in \gamma(0; Q)^*$ and $|w/z|<1$ for $w \in \gamma(0; P)^*$. We now invoke Theorem 3.13 [uniform convergence] in order to interchange summation and integration; cf. the proof of Theorem 5.9. This gives

$$f(z) = \sum_{n=0}^{\infty} \left(\frac{1}{2\pi i} \int_{\gamma(0; Q)} \frac{f(w)}{w^{n+1}} \, dw \right) z^n$$

$$+ \sum_{m=0}^{\infty} \left(\frac{1}{2\pi i} \int_{\gamma(0; P)} f(w) w^m \, dw \right) z^{-m-1}.$$

Putting $n = -m - 1$ in the second sum and using the Deformation theorem to replace $\gamma(0; Q)$ and $\gamma(0; P)$ by γ (as specified in the statement of Laurent's theorem), we have the required expansion for $f(z)$. $\square$

6.2 Uniqueness of the Laurent expansion

Let $f \in H(A)$, where $A = \{z : R < |z - a| < S\}$ $(0 \leq R < S \leq \infty)$, and suppose

$$f(z) = \sum_{n=-\infty}^{\infty} b_n (z - a)^n \quad (z \in A).$$

Then $b_n = c_n$ for all n, where c_n is as in Theorem 6.1.

Proof. We may assume $a = 0$. Choose r such that $R < r < S$. Then

$$2\pi i c_n = \int_{\gamma(0; r)} f(w) w^{-n-1} \, dw = \int_{\gamma(0; r)} \sum_{k=-\infty}^{\infty} b_k w^{k-n-1} \, dw$$

$$= \int_{\gamma(0; r)} \sum_{k=0}^{\infty} b_k w^{k-n-1} \, dw + \int_{\gamma(0; r)} \sum_{m=1}^{\infty} b_{-m} w^{-m-1-n} \, dw.$$

Theorem 3.13 applies to each of these integrals. Hence summation and integration can be interchanged to give

$$2\pi i c_n = \sum_{k=-\infty}^{\infty} b_k \int_{\gamma(0; r)} w^{k-n-1} \, dw = 2\pi i b_n \quad \text{(by 3.7(1)).} \quad \square$$

6.3 Computation of Laurent expansions

Thanks to Corollary 2.13, coefficients in Taylor expansions can be computed quite easily. There is no such handy method for finding the Laurent coefficients

$$c_n = \frac{1}{2\pi i} \int_{\gamma} \frac{f(w)}{(w - a)^{n+1}} \, dw,$$

and these would be tedious, or impossible, to work out from first principles. Fortunately the uniqueness theorem, 6.2, comes to the rescue: it tells us that if, by hook or by crook, we can find a valid expansion of Laurent type for a function holomorphic in an annulus, then the coefficients in the expansion are the official Laurent coefficients c_n. This means that we can combine simple known expansions (Taylor or Laurent) to obtain Laurent expansions of quite complicated functions. In applications often only the first few terms in the expansion are required.

6.4 Examples

(1) $f(z) = 1/[z(1-z)]$ is holomorphic in each of the annuli: $A_1 = \{z : 0 < |z| < 1\}$, $A_2 = \{z : |z| > 1\}$. We have

$$f(z) = z^{-1} + (1-z)^{-1},$$

so

$$f(z) = \sum_{n=-1}^{\infty} z^n \qquad (z \in A_1) \qquad (*)$$

and

$$f(z) = z^{-1} - z^{-1}(1 - z^{-1})^{-1} = \sum_{n=-\infty}^{-2} -z^n \qquad (z \in A_2) \qquad (†)$$

(by the binomial expansions for $(1-z)^{-1}$ for $|z| < 1$ and $|z| > 1$ (recall 2.8)). Uniqueness implies that (*) and (†) give the Laurent expansions for $f(z)$ in A_1 and A_2 respectively.

(2) The function $f(z) = \dfrac{1}{z(1-z)^2}$ is holomorphic for $0 < |z-1| < 1$; for z in this annulus, the binomial expansion gives

$$f(z) = \frac{1}{(z-1)^2}\left(\frac{1}{1+(z-1)}\right) = \frac{1}{(z-1)^2}[1 - (z-1) + (z-1)^2 - \ldots].$$

So the Laurent expansion for $f(z)$, valid for $0 < |z-1| < 1$, is

$$f(z) = \sum_{n=-2}^{\infty} (-1)^n (z-1)^n.$$

(3) The function cosec z is holomorphic except at $z = k\pi$ $(k \in \mathbb{Z})$ and so has a Laurent expansion $\sum_{n=-\infty}^{\infty} c_n z^n$ valid for $0 < |z| < \pi$. We have

$$\sin z := z - \frac{z^3}{3!} + \frac{z^5}{5!} - \ldots = z\left(1 - \frac{z^2}{3!} + h(z)\right),$$

where all the terms after the first two have been amalgamated to form the holomorphic function h. Near 0, $|h(z)| \leqslant K |z|^4$ for some constant K; we shall use the conventional O-notation for this, and write $h(z) = O(z^4)$. Then

$$\operatorname{cosec} z = \frac{1}{z}\left[1 - \left(\frac{z^2}{3!} + O(z^4)\right)\right]^{-1} = \frac{1}{z}\left(1 + \frac{z^2}{3!} + O(z^4)\right) \quad \text{for small } |z|$$

(using $(1 - w)^{-1} = 1 + w + w^2 + \ldots$, which is valid for $|w| < 1$, with $w = z^2/3! + h(z)$). Then 6.2 implies $c_n = 0$ for all $n < -1$, $c_{-1} = 1$, $c_1 = 1/6$. By taking more terms in the above expansions, we could compute c_3, c_5, $\ldots$. Trivially $c_{2k} = 0$ for all integers k.

To find the Laurent expansion of $\operatorname{cosec} z$ about $z = k\pi$ ($k \in \mathbb{Z}$, $k \neq 0$), note that

$$\operatorname{cosec} z = (-1)^k \operatorname{cosec}(z - k\pi) = \sum_{n=-\infty}^{\infty} (-1)^k c_n (z - k\pi)^n,$$

where the coefficients c_n are as in the expansion about zero.

(4) $\cot z = \dfrac{\cos z}{\sin z}$ is holomorphic for $0 < |z| < \pi$. Near 0,

$$\cot z = \left(1 - \frac{z^2}{2!} + O(z^4)\right)\left(\frac{1}{z} + \frac{z}{6} + O(z^3)\right) \quad \text{(by (3))}$$

$$= \frac{1}{z}\left(1 + z^2\left(-\frac{1}{2} + \frac{1}{6}\right) + O(z^4)\right),$$

since multiplication of convergent power series is permissible, by 5.11. Hence

$$\cot z = \frac{1}{z} - \frac{z}{3} + O(z^3) \quad (0 < |z| < \pi).$$

(5) Our last example in this group is a more theoretical one. It illustrates how the formula for the Laurent coefficients of a function f can be used to translate information about f into information about its Laurent series. Suppose that f is holomorphic in $\{z : |z| > R\}$ and that f is bounded ($|f(z)| \leqslant M$, say). We have, for $|z| > R$,

$$f(z) = \sum_{n=-\infty}^{\infty} c_n z^n, \quad \text{where } c_n = \frac{1}{2\pi i} \int_{\gamma(0;r)} \frac{f(w)}{w^{n+1}} \, dw.$$

Intuition suggests that the boundedness of f forces $c_n = 0$ for $n > 0$. We prove this by estimating c_n (cf. 5.10). By 3.10,

$$|c_n| \leqslant \frac{1}{2\pi} \sup\{|f(w)w^{-(n+1)}| : |w| = r\} \times 2\pi r \leqslant Mr^{-n}.$$

This holds for all $r > R$. Since $r^{-n} \to \infty$ as $r \to \infty$, for $n > 0$, we conclude that $c_n = 0$ for $n > 0$. Hence f has the expansion

$$f(z) = \sum_{n=-\infty}^{0} c_n z^n \qquad (|z| > R).$$

Singularities

6.5 Definitions

Let f be a complex-valued function. The point a is a *regular point* if f is holomorphic at a (that is, if there exists r such that $f \in H(D(a\,;r))$; see 2.2(3)). The point a is a *singularity* of f if a is a limit point of regular points which is not itself regular.

   If a is a singularity of f and f is holomorphic in some punctured disc $D'(a\,;r)$, then a is an *isolated singularity*; if $f \notin H(D'(a\,;r))$ for any $r > 0$, a is a *non-isolated essential singularity*.

6.6 Classification of isolated singularities

Suppose f has an isolated singularity at a. Then f is holomorphic in some annulus $\{z : 0 < |z - a| < r\}$ and there has a *unique* Laurent expansion

$$f(z) = \sum_{n=-\infty}^{\infty} c_n (z - a)^n.$$

The point a is said to be:
   a removable singularity if $c_n = 0$ for all $n < 0$;
   a pole of order m $(m \geqslant 1)$ if $c_{-m} \neq 0$ and $c_n = 0$ for all $n < -m$;
   an isolated essential singularity if there does not exist m such that $c_n = 0$ for all $n < -m$.
Poles of orders 1, 2, 3, ... are called *simple, double, triple,* ....

Notes (1) Uniqueness of the Laurent coefficients ensures that these definitions make sense.
(2) In $D'(a\,;r)$,

$$f(z) = \sum_{n=-\infty}^{-1} c_n (z - a)^n + \sum_{n=0}^{\infty} c_n (z - a)^n.$$

The first sum on the right-hand side is the *principal part* of the Laurent expansion; the second sum is holomorphic in $D(a\,;r)$, by 2.12. Notice that $f(z) - \sum_{n=-\infty}^{-1} c_n (z - a)^n$ has a removable singularity at a. For more information on removable singularities see 6.12(1).

6.7 Examples

(1) $(z - 1)^{-2}$ has a double pole at $z = 1$.
(2) $(1 - \cos z) z^{-2}$ is holomorphic except at $z = 0$, where it is inde-

terminate. The Laurent expansion about $z = 0$ is

$$\frac{1}{2} - \frac{z^2}{4!} + \frac{z^4}{6!} - \cdots,$$

so the singularity at 0 is removable.

(3) We showed in Example 6.4(4) that

$$\cot z = \frac{1}{z} - \frac{z}{3} + O(z^3) \qquad (z \in D'(0; \pi)).$$

Hence $\cot z$ has a simple pole at 0. Since $\cot(z - k\pi) = \cot z$ for each integer k, each singularity $k\pi$ of $\cot z$ is a simple pole.

(4) If $0 < |z| < \infty$,

$$\sin\left(\frac{1}{z}\right) = \sum_{n=0}^{\infty} (-1)^n \frac{z^{-(2n+1)}}{(2n+1)!}.$$

Hence $\sin(1/z)$ has an isolated essential singularity at 0.

(5) $\mathrm{cosec}(1/z)$ has singularities at $1/(k\pi)$ $(k \in \mathbb{Z})$. For $k \neq 0$ there is a simple pole at $1/(k\pi)$. Since $\mathrm{cosec}(1/z)$ is not holomorphic in any punctured disc $D'(0; r)$, the point 0 is *not* an isolated singularity. See also 6.15.

It should be clear from the examples in 6.4 that direct computation of the Laurent coefficients is an arduous way of classifying the singularities of even relatively simple functions. The clue to a more efficient method lies in the observation that, if a holomorphic function has an isolated zero at the point a, then its reciprocal has an isolated singularity at a. To exploit this to the full we need some preliminary facts about zeros, and a technical theorem.

6.8 Zeros

Suppose that $f \in H(D(a ; r))$ for some r and that $f(a) = 0$. Assume that f is not identically zero in $D(a; r)$. By Taylor's theorem, 5.9,

$$f(z) = \sum_{n=m}^{\infty} c_n(z - a)^n \qquad (z \in D(a; r)), \quad \text{where} \quad m \geq 1 \text{ and } c_m \neq 0.$$

We define the *order* of the zero of f at a to be m. Zeros of orders. 1, 2, ... are called *simple, double,* .... Since, by 2.13, $c_n = f^{(n)}(a)/n!$, f has a zero of order m at a if and only if

$$f(a) = f'(a) = \ldots = f^{(m-1)}(a) = 0, \qquad f^{(m)}(a) \neq 0.$$

6.9 Theorem

(1) Let $f \in H(D(a ; r))$. Then f has a zero of order m at a if and only if

$$\lim_{z \to a}(z - a)^{-m}f(z) = C, \text{ where } C \text{ is a non-zero constant.} \qquad (*)$$

(2) Let $f \in H(D'(a\,;r))$. Then f has a pole of order m at a if and only if

$$\lim_{z \to a}(z-a)^m f(z) = D, \text{ where } D \text{ is a non-zero constant.} \qquad (\dagger)$$

Proof. We prove (2). The proof of (1) is very similar, and is left as an exercise.

Necessity Suppose a is a pole of order m. For $z \in D'(a\,;r)$,

$$f(z) = \sum_{n=-m}^{\infty} c_n(z-a)^n, \quad \text{where} \quad c_{-m} \neq 0.$$

In $D'(a\,;r)$, $(z-a)^m f(z) = \sum_{n=0}^{\infty} c_{n-m}(z-a)^n$. The series on the right-hand side defines a function continuous at $z = a$ (by 2.12 and 2.3(1)). Therefore

$$\lim_{z \to a}(z-a)^m f(z) = c_{-m} \neq 0.$$

Sufficiency By Laurent's theorem, 6.1,

$$f(z) = \sum_{n=-\infty}^{\infty} c_n(z-a)^n, \quad \text{where}$$

$$c_n = \frac{1}{2\pi i} \int_{\gamma(a\,;s)} \frac{f(w)}{(w-a)^{n+1}}\, dw \quad (0 < s < r).$$

We require $c_n = 0$ $(n < -m)$, $c_{-m} \neq 0$. By the condition $(\dagger)$, given $\varepsilon > 0$, there exists $\delta > 0$ such that

$$|(w-a)^m f(w) - D| < \varepsilon \quad \text{whenever} \quad 0 < |w-a| < \delta.$$

Take $0 < s < \min\{\delta, r\}$. Then

$$|w-a| = s \Rightarrow |(w-a)^m f(w)| \leqslant |D| + \varepsilon \quad \text{(by 1.4(2))}$$
$$\Rightarrow |(w-a)^{-n-1} f(w)| \leqslant (|D| + \varepsilon)s^{-n-m-1}.$$

Hence, by estimating the integral defining c_n (using 3.10)),

$$|c_n| \leqslant (|D| + \varepsilon)s^{-n-m}.$$

If $n < -m$, s^{-n-m} can be made arbitrarily small by taking s sufficiently small. The constant c_n is independent of s, so $c_n = 0$.

We now have $f(z) = \sum_{n=-m}^{\infty} c_n(z-a)^n$. As in the necessity proof,

$$c_{-m} = \lim_{z \to a}(z-a)^m f(z) = D \neq 0. \qquad \square$$

6.10 Corollaries

(1) Suppose f is holomorphic in some disc $D(a\,;r)$. Then f has a zero of order m at a if and only if $1/f$ has a pole of order m at a.

(2) Suppose that f has a pole of order m at a.

(i) Suppose $g \in H(D(a\,;r))$ for some r. Then, at a, the function fg has

a pole of order m if $g(a) \neq 0$,

a pole of order $m - n$ if g has a zero of order n at a and $n < m$,

a removable singularity if g has a zero of order at least m.

(ii) Suppose g has a pole of order n at a. Then fg has a pole of order $m + n$ at a.

Proof. (1) Suppose $1/f$ has a pole at a. Then f cannot have a non-isolated zero at a, and so, by the Identity theorem, 5.14, f is non-zero in $D'(a;s)$ for some $s > 0$. The result (sufficiency and necessity) now follows from Theorem 6.9.

The proof of (2) is left as an exercise. $\qquad\qquad\qquad\square$

6.11 Examples

(1) $z \sin z$ has zeros at $z = n\pi$ $(n \in \mathbb{Z})$.

$$\left[\frac{d}{dz}(z \sin z)\right]_{z=0} = \left[(\sin z + z \cos z)\right]_{z=0} = 0, \quad \left[\frac{d^2}{dz^2}(z \sin z)\right]_{z=0} \neq 0,$$

and for $n \neq 0$,

$$\left[\frac{d}{dz}(z \sin z)\right]_{z=n\pi} \neq 0.$$

Hence, by 6.8 and 6.10, $1/(z \sin z)$ has a double pole at 0 and simple poles at $z = n\pi$ $(n \in \mathbb{Z}, n \neq 0)$.

(2) Consider $\cot z$. At the points $n\pi$ $(n \in \mathbb{Z})$, $\sin z$ has simple zeros (by 6.8) and $\cos z \neq 0$. Hence, by 6.10, $\cot z$ has simple poles at $z = n\pi$ $(n \in \mathbb{Z})$. Compare this method with that of 6.7(3).

(3) Consider

$$F(z) = \frac{(z-1)^2 \cos \pi z}{(2z-1)(z^2+1)^5 \sin^3 \pi z}.$$

The denominator has a simple zero at $1/2$, zeros of order 5 at $\pm i$ and a triple zero at k for each $k \in \mathbb{Z}$; the numerator has a double zero at 1 and simple zero at $(2k+1)/2$ for each $k \in \mathbb{Z}$. Appealing to 6.10 we see that f has poles of order 5 at $\pm i$, a simple pole at 1, a triple pole at k, for $k \in \mathbb{Z}$, $k \neq 1$, and a removable singularity at $1/2$.

6.12 Behaviour near an isolated singularity

(1) **Removable singularity** Suppose $f(z) = \sum_{n=0}^{\infty} c_n(z-a)^n$ in $D'(a;r)$. Then $f(z) \to c_0$ as $z \to a$. By defining (or redefining) $f(a)$

to be c_0, we make

$$f(z) = \sum_{n=0}^{\infty} c_n(z-a)^n \quad \text{in } D(a;r),$$

and so make f holomorphic in $D(a;r)$, by 2.12. Thus a removable singularity is something of a non-event: a ceases to be classified as a singularity once f is correctly defined at a.

(2) **Pole** Suppose f has a pole at a. It is immediate from Corollary 6.10(1) that $|f(z)| \to \infty$ as $z \to a$.

(3) **Essential singularity** Suppose f has an isolated essential singularity at a. Let w be *any* complex number. Then there exists a sequence $\langle a_n \rangle$ such that $a_n \to a$ and $f(a_n) \to w$. This is the *Casorati–Weierstrass theorem*. For an outline of the proof, see Exercise 6.10.

A more spectacular and much deeper result, due to Picard, asserts that in any $D'(a;r)$, f actually assumes every complex value, except possibly one. In the case of $e^{1/z}$, which has an isolated essential singularity at 0, the exception is 0.

Meromorphic functions

6.13 The extended complex plane

For the analysis of the behaviour of holomorphic functions as $|z| \to \infty$ we introduce a convenient device, the *extended complex plane* $\tilde{C}$. We form $\tilde{C}$ by adding an extra point ∞ to C. We can give our construction geometrical significance as follows: we regard C as embedded in Euclidean space $\mathbb{R}^3$ by identifying $z = x + iy$ with $(x, y, 0)$. We let

$$\Sigma := \{(x, y, u) \in \mathbb{R}^3 : x^2 + y^2 + (u - \tfrac{1}{2})^2 = \tfrac{1}{4}\};$$

this is a sphere (the *Riemann sphere*), touching the plane C at $(0, 0, 0)$. Stereographic projection (see Fig. 6.2) allows us to set up a one-to-one correspondence between $\tilde{C}$ and Σ, under which

$$C \ni z = x + iy = re^{i\theta} \leftrightarrow z = (x(1 + r^2)^{-1}, y(1 + r^2)^{-1}, r^2(1 + r^2)^{-1}),$$

$$\infty \leftrightarrow (0, 0, 1), \text{ the north pole of } \Sigma.$$

We define open discs in $\tilde{C}$ as follows:

$$D(a;r) := \{z \in C : |z - a| < r\} \quad (a \in C, r > 0),$$
$$D(\infty;r) := \{z \in C : |z| > r\} \cup \{\infty\} \quad (r > 0).$$

A subset G of $\tilde{C}$ is said to be *open* if, given $z \in G$, there exists r

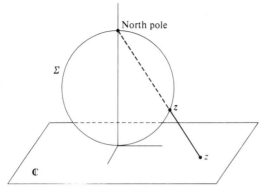

North pole

Σ

$\mathbb{C}$

Fig. 6.2

such that $D(z; r) \subseteq G$. [The open sets in $\tilde{\mathbb{C}}$ correspond to the intersections with Σ of open spheres in $\mathbb{R}^3$. We have defined a topology on $\tilde{\mathbb{C}}$ so as to make $\tilde{\mathbb{C}}$ homeomorphic to the sphere Σ. As a closed bounded subset of $\mathbb{R}^3$, Σ is compact, so $\tilde{\mathbb{C}}$ is compact, and can be regarded as a one-point compactification of the non-compact space $\mathbb{C}$.]

Continuity, etc., of functions on $\tilde{\mathbb{C}}$ can be handled topologically. More naively, we may decree that a complex-valued function f defined on a set containing some disc $D(\infty; r)$ is *continuous* (*differentiable*) at ∞ if and only if $\tilde{f}$ defined on $D(0; 1/r) \subseteq \mathbb{C}$ by

$$\tilde{f}(z) = f(1/z) \ (z \neq 0), \qquad \tilde{f}(0) = f(\infty)$$

is continuous (differentiable) at 0. Similarly, all the terms relating to zeros and singularities are applied to ∞. For example, f is said to have a *pole at* ∞ if and only if $\tilde{f}$ has a pole at 0; e.g. at ∞, z^3 has a triple pole and $1/z^2 \sin(1/z)$ a removable singularity.

6.14 Definition

Let $G \subseteq \tilde{\mathbb{C}}$ be open. A complex-valued function which is holomorphic in G except possibly for poles is said to be *meromorphic* in G.

6.15 Limit points of singularities

Closed sets and limit points in $\tilde{\mathbb{C}}$ are defined in the same way as in $\mathbb{C}$; see 1.10. We claim that any infinite closed subset S of $\tilde{\mathbb{C}}$ has a limit point in S. We prove this from the Bolzano–Weierstrass theorem as presented in 1.17. If there exists R such that $S \subseteq \bar{D}(0; R)$, our claim follows immediately from 1.17. Otherwise, $S \cap \{z \in \mathbb{C} : |z| > r\} \neq \varnothing$ for any r, in which case ∞ is the required limit point. [The claim can be established more elegantly by invoking the Bolzano–Weierstrass theorem for Σ.]

Let a be a limit point of singularities of a function f defined on some subset of $\tilde{\mathbb{C}}$. Then f cannot be holomorphic in any punctured disc $D'(a; r)$, and cannot have a Laurent expansion about a. Hence a is neither a regular point nor an isolated singularity. Thus a limit point of singularities is a non-isolated essential singularity.

Consider, for example, $f(z) = z^{-3} \sec(1/z)$. The factor $\sec(1/z)$ is undefined at 0, and has simple poles at the points $2/[(2n+1)\pi]$ $(n \in \mathbb{Z})$ where $\cos(1/z)$ has simple zeros. Hence 0 is a limit point of poles of f, and so a non-isolated essential singularity. The fact that the factor z^{-3} has a triple pole at 0 is a red herring; f itself cannot have a Laurent expansion about 0.

Suppose that f is meromorphic in an open set $G \subseteq \tilde{\mathbb{C}}$. Then, by the observations preceding the example,

(i) the set of poles of f has no limit point in G, and so

(ii) f can have at most a finite number of poles in any closed set contained in G.

The reward for our efforts in extending topological concepts to $\tilde{\mathbb{C}}$ is reaped in Theorem 6.16, which is elegantly simple. For a concrete instance of its proof, see Example 7.5.

6.16 Theorem

(1) Let f be holomorphic in $\tilde{\mathbb{C}}$. Then f is constant.
(2) Let f be meromorphic in $\tilde{\mathbb{C}}$. Then f is a rational function.

Proof. (1) The result follows from Liouville's theorem, 5.2, once we know f is bounded [which, as a continuous function on a compact space, it is]. An elementary proof of boundedness goes as follows. The function $\tilde{f}$ (defined by $\tilde{f}(z) = f(1/z)$ $(z \neq 0)$, $\tilde{f}(0) = f(\infty)$) is continuous on the compact set $\bar{D}(0; 1)$ in $\mathbb{C}$, and hence is bounded there, by 1.18. Consequently f is bounded on $\{z : |z| \geq 1\} \cup \{\infty\}$. By 1.18 again, f is bounded on $\{z : |z| \leq 1\}$.

(2) By 6.15, f has at most a finite number of poles. Assume f has a pole of order m_k at a_k $(k = 1, \ldots, N)$. Let the principal part of the Laurent expansion of f about a_k be

$$f_k(z) := \sum_{n=-m_k}^{-1} c_{kn}(z - a_k)^n \qquad (k = 1, \ldots, N),$$

and about ∞ be $f_0(z)$, so that, if f has a pole at infinity,

$$f_0(z) := \sum_{n=1}^{m} c_{0n} z^n, \quad \text{for some } m \geq 1.$$

Then

$$g(z) := f(z) - \sum_{k=0}^{N} f_k(z)$$

defines a function g with only removable singularities (see the note in 6.6); remove them (see 6.12(1)) to obtain a function holomorphic in $\tilde{\mathbb{C}}$. This function is constant, so f is a rational function. $\square$

Multifunctions

This section deals with the ticklish subject of multifunctions, in a way suited to the solution of later problems involving the evaluation of integrals and conformal mapping. Since this section is likely to be needed by Level I readers, we have not presumed familiarity with the material on logarithms in Chapter 4. The rules we lay down for handling multifunctions can be mastered quite easily. However the theory underlying them essentially relies on the section in Chapter 4 on logarithms, argument, and index.

6.17 Introductory remarks

We recall that a multifunction w defined on a set $S \subseteq \mathbb{C}$ is an assignment to each $z \in S$ of a set $[w(z)]$ of complex numbers. We shall concentrate on logarithms, powers and roots of rational functions; for the definitions, see 2.19 and 2.20. A function f on S assigns to each $z \in S$ a unique complex number $f(z)$; in this section we shall often for emphasis refer (tautologically) to a 1-valued function.

We pose the following problem: given a multifunction w defined on S, can we select, for each $z \in S$, a point $f(z)$ in $[w(z)]$ so that f is holomorphic in an open subset G of S, where G is to be chosen as large as possible? If we are to do this, then $f(z)$ must vary continuously with z in G (since a holomorphic function is necessarily continuous). Suppose w is defined in some $D'(a; R)$ and that $f(z) \in [w(z)]$ is chosen so that f is, at least, continuous on the circle with centre a and radius r ($0 < r < R$). As z traces out $\gamma(a; r)^*$ starting from, say, ζ, $f(z)$ varies continuously, but must be restored to its original value $f(\zeta)$ when z completes its circuit, because f is, by hypothesis, 1-valued. Notice also that if $z - a = re^{i\theta(z)}$, where $\theta(z)$ is chosen to vary continuously with z, then $\theta(z)$ increases by 2π as z performs its circuit (see 4.19(1)), so that $\theta(z)$ is *not* restored to its original value. The same phenomenon does not occur if z moves round a circle in $D'(a; R)$ not containing a: in this case $\theta(z)$ does return to its original value (see 4.19(1)).

Now take the logarithm as an example. If

$$f(z) \in [\log z] := \{\log|z| + i\theta : \theta \in [\arg z]\}$$

is to be chosen so that f is continuous on the circle $\gamma(0; r)^*$, then θ must be selected to vary continuously with z. Our previous observation shows this to be incompatible with 1-valuedness of f. This indicates [as we would expect from Chapter 4] that we cannot choose f to yield a holomorphic logarithm in any open set which encircles 0. More generally, our discussion suggests that if we are to extract a holomorphic function from a multifunction w, we shall need to restrict to a set in which it is impossible to encircle, one at a time, points a such that the definition of $[w(z)]$ involves the argument of $z - a$. (As we shall see in Example 6.20(3), encircling several of these 'bad' points simultaneously may be allowable.) Before proceeding with generalities, we turn to the logarithm to provide further motivation.

6.18 The logarithm

We have $[\log z] = \{\log|z| + i\theta : \theta \in [\arg z]\}$ $(z \neq 0)$. The many-valuedness of the logarithm occurs because for any given z there are infinitely many choices for $\theta \in [\arg z]$. We can define an associated family of 1-valued functions as follows. For $k \in \mathbb{Z}$,

$$F_k(r, \theta) = \log r + i(\theta + 2k\pi) \qquad (r > 0, \, \theta \in \mathbb{R}).$$

Each F_k is a continuous function of r and θ. The functions F_k can be used in two ways to generate, without repetitions, all the values in $[\log z_0]$ for a given $z_0 \neq 0$. We have

(a) $[\log z_0] = \{F_{k_0}(|z_0|, \theta) : \theta \in [\arg z]\}$, for any fixed $k_0 \in \mathbb{Z}$, and
(b) $[\log z_0] = \{F_k(|z_0|, \theta_0) : k \in \mathbb{Z}\}$, for a fixed representation $|z_0|e^{i\theta_0}$ of z_0.

We stress that we cannot, as we might naively have hoped, use F_k to define a 1-valued logarithm in $\mathbb{C}\setminus\{0\}$ simply by taking

$$w_k(z) = w_k(re^{i\theta}) = F_k(r, \theta) \qquad (0 \neq z = re^{i\theta});$$

as (a) indicates, w_k so defined produces a multifunction. However, the functions F_k (which we shall call multibranches) are useful for tracking what happens when we select continuously from $[\log z]$ as z moves.

Let γ be a circle containing 0 and let z trace out γ^*, starting at, and returning to, ζ. Write $z = |z|e^{i\theta}$, where θ is chosen to vary continuously with z. Let the initial value of θ be ϕ. Then the final

value is $\phi + 2\pi$. Suppose our initial determination of the logarithm is $F_k(|\zeta|, \phi)$. If we select values from $[\log z]$ continuously as z moves, then, having started with F_k, we must stick with it. To switch to F_m ($m \neq k$) en route would cause a discontinuity. Thus the initial choice from $[\log \zeta]$ is $F_k(|\zeta|, \phi)$, and the final value (forced by continuous choice) is $F_k(|\zeta|, \phi + 2\pi) = F_{k+1}(|\zeta|, \phi)$. Hence we can think of the circuit as inducing a permutation of the multibranches F_k. In symbols:

$$F_k \underset{\gamma}{\to} F_{k+1} \qquad (k \in \mathbb{Z}).$$

If z traced out a circle not containing 0, no such non-trivial permutation of the multibranches would occur. All this just confirms what we said in 6.17: we cannot select a continuous 1-valued logarithm unless encirclement of 0 is forbidden. One way to prevent z circling round 0 is to imagine a cut made in the plane along some infinite half-line from 0, and in the cut plane to outlaw paths which cross the cut. We opt for a cut along the negative real axis. We may think of this cut as having two 'edges', the 'upper' edge being given by $z = re^{i\pi}$ and the 'lower' by $z = re^{-i\pi}$ ($r > 0$). For $k \in \mathbb{Z}$, let $f_k(z) = F_k(r, \theta) = \log r + i(\theta + 2k\pi)$, where $0 \neq z = re^{i\theta}$ and $-\pi < \theta \leqslant \pi$. The restriction on θ ensures that each f_k is a 1-valued function in $\mathbb{C} \backslash \{0\}$. At points of the cut we have elected to use 'upper edge values'. Each f_k is continuous, except, as is inevitable, across the cut. In crossing the cut from the upper half-plane to the lower half-plane we transfer continuously from f_k to f_{k+1}.

Each f_k is holomorphic in $\mathbb{C}_\pi = \mathbb{C} \backslash (-\infty, 0]$, with $f_{k'}(z) = 1/z$. To see this, write $\zeta = f_k(z)$ and $\eta = f_k(z + h) - f_k(z)$. By continuity of f_k, $\eta \to 0$ as $h \to 0$. Hence

$$[f_k(z + h) - f_k(z)]/h = \eta/(e^{\zeta + \eta} - e^\zeta) \to 1/e^\zeta \text{ as } h \to 0.$$

[In the simply connected region $\mathbb{C}_\pi$ (which is just the plane with the points of the cut excluded), the functions f_k are just the different possible holomorphic logarithms; see 4.15 and Exercise 4.6.] We shall call the functions f_k holomorphic branches of the logarithm, and the point 0 which caused all the trouble a branch point.

6.19 Branch points, multibranches, cuts, and branches

Notwithstanding our preparatory treatment of the logarithm, this subsection, which sets up the vocabulary for discussing multifunctions, may seem somewhat abstruse at first. Reference back to 6.17 and 6.18 and forward to the examples in 6.20 should make the ideas easier to understand. To attempt to define such notions as branch point and branch for too wide a class of multifunctions would be foolhardy, and what follows should be taken to apply to:

 (i) the logarithm, with domain of definition $S = \mathbb{C}\backslash\{0\}$;

 (ii) the general power, with domain of definition $S = \mathbb{C}\backslash\{0\}$;

 (iii) the logarithm of, or a power of, a non-constant rational function $p(z)/q(z)$, defined except on the set of zeros of p and of q.

This list could be extended somewhat, but it is adequate for practical purposes.

 Suppose w is a multifunction defined on S. A point $a \in \mathbb{C}$ is said to be a *branch point* of w if, for all sufficiently small $r > 0$, it is not possible to choose $f(z) \in [w(z)]$ so that f defines a continuous function on $\gamma(a; r)^*$. We say ∞ is a branch point of w if 0 is a branch point of $\tilde{w}$, where $[\tilde{w}(z)] := [w(1/z)]$ ($z \in S, z \neq 0$). It turns out that for a multifunction of any of the types mentioned above the branch points in $\mathbb{C}$ are exactly the points we have excluded from the domain of definition, and that ∞ is sometimes a branch point but not always. We shall prove this in specific cases below.

 We have already defined multibranches for the logarithm. We shall define multibranches for other multifunctions individually in due course. Here we simply list the properties our multibranches have. Let w be a multifunction defined on $S = \mathbb{C}\backslash\{a_1, \ldots, a_s\}$. For each $z \in S$, let $\mathbf{r}(z) := (r_1, \ldots, r_s)$, where $r_j = |z - a_j|$ ($1 \leqslant j \leqslant s$) and

$$\mathbf{\Theta}(z) := \{\mathbf{\theta} := (\theta_1, \ldots, \theta_s) : \theta_j \in [\arg(z - a_j)] \quad (1 \leqslant j \leqslant s)\}.$$

A set of multibranches for w is a set $\{F_k : k \in K\}$ of functions indexed by a (finite or infinite) subset K of $\mathbb{Z}$ such that

 (i) each F_k is a continuous complex-valued function on the product set $\mathbb{R}_+^s \times \mathbb{R}^s$ (where $\mathbb{R}_+ := \{x \in \mathbb{R} : x > 0\}$);

 (ii) $\{F_k : k \in K\}$ is a minimal set of such functions sufficient to generate the values of w at any given $z_0 \in S$ in two ways:

 (a) for each fixed $k_0 \in K$,

$$[w(z_0)] = \{F_{k_0}(\mathbf{r}(z_0), \mathbf{\theta}) : \mathbf{\theta} \in \mathbf{\Theta}(z_0)\}.$$

 (b) for each fixed choice of $\mathbf{\theta}_0 \in \mathbf{\Theta}(z_0)$,

$$[w(z_0)] = \{F_k(\mathbf{r}(z_0), \mathbf{\theta}_0) : k \in K\};$$

 Assume a set $\{F_k : k \in K\}$ of multibranches has been constructed for w. Now let γ be any contour in S and allow z to trace out γ^*, starting and finishing at ζ. For each $j = 1, \ldots, s$, let $\theta_j \in [\arg(z - a_j)]$ be chosen to vary continuously as z moves round γ^*; see Fig. 6.3. Let the initial value of θ_j be ϕ_j and let the value of θ_j when z has

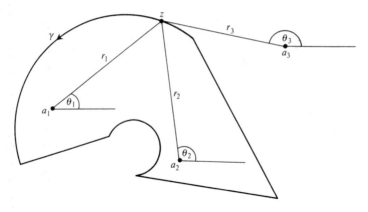

Fig. 6.3

completed its circuit be Φ_j. (So $\Phi_j = \phi_j + 2\pi$ if a_j is inside γ and $\Phi_j = \phi_j$ otherwise.) Condition (ii) implies that $\{F_k(\mathbf{r}(\zeta), \boldsymbol{\phi}) : k \in K\}$ and $\{F_k(\mathbf{r}(\zeta), \boldsymbol{\Phi}) : k \in K\}$ contain the same points; each consists of the set of values of w at ζ. However it may be that the values are permuted: we may have

$$F_k(\mathbf{r}(\zeta), \boldsymbol{\Phi}) = F_m(\mathbf{r}(\zeta), \boldsymbol{\phi}) \quad \text{for some } k \text{ and } m \text{ with } k \neq m.$$

We call γ 'bad' if this happens, so that γ is 'bad' if a circuit round γ^* induces a non-trivial permutation of the multibranches. Any continuous selection of values from $[w(z)]$ as z moves round γ^* must be made using a single F_k, given that $\theta_1, \ldots, \theta_s$ are chosen continuously. Consequently multibranches can also be used to identify branch points: a point $a \in \mathbb{C}$ is a branch point if and only if $\gamma(a; r)$ is 'bad' for all sufficiently small r.

Having identified the branch points and the 'bad' contours, we make a minimal set of cuts between branch points, along non-intersecting polygonal lines and infinite half-lines, so that every 'bad' contour crosses some cut. If we make it a rule that in a cut plane a contour is forbidden to cross any cut, we outlaw 'bad' contours. Denote by $\mathbb{C}_{\text{cut}}$ the plane with the points of the cuts removed. It is possible to choose for each j and for each $z \in \mathbb{C}_{\text{cut}}$, $\theta_j(z) \in [\arg(z - a_j)]$ such that $\theta_j(z)$ is a 1-valued continuous function of z in $\mathbb{C}_{\text{cut}}$. [A formal proof of this involves Theorem 4.15.] We can now define for each $k \in K$ a 1-valued continuous function f_k of z in $\mathbb{C}_{\text{cut}}$ by

$$f_k(z) = F_k(\mathbf{r}(z), \boldsymbol{\theta}(z)), \quad \text{where} \quad \boldsymbol{\theta}(z) = (\theta_1(z), \ldots, \theta_s(z)).$$

We shall, as is customary in the special case of the logarithm, also assign values of f_k on the cuts, except at branch points. This we do by taking the limiting values obtained by approaching each cut from a specified side; such an extension to all of S is, of course, not unique (but this matters little). The functions f_k are called *branches*. A branch is said to be *holomorphic* if it is holomorphic at each point of the associated cut plane $\mathbb{C}_{cut}$. It turns out that for the class of multifunctions we are considering branches are always holomorphic. This is because of the way they arise as inverses of holomorphic functions; the last part of the proof of Theorem 10.25 gives us just what we want, since branches are continuous.

Notice that the position of cuts is far from unique. Consider, for example, the logarithm. We already know that 0 is a branch point, and since $[\log(1/z)] = [-\log z]$, ∞ is also a branch point. A cut along any half-line from 0 to ∞ will serve. Notice also that the way branches are defined is linked to the positions of the cuts. With the logarithm, had we chosen alternatively to cut the plane along $[0, \infty)$, we should have taken branches f_k given by

$$f_k(z) = \log r + i(\theta + 2k\pi) \quad (0 \neq z = re^{i\theta}, 0 \leq \theta < 2\pi);$$

as before, 'upper edge values' have been chosen on the cut.

6.20 Examples

(1) **Powers** Let $\alpha \notin \mathbb{Z}$. Then $[z^\alpha]$ is really just $[e^{\alpha \log z}]$; see 2.20. The branch points are 0 and ∞, and we cut the plane along $(-\infty, 0]$. Each holomorphic branch f_k of the logarithm (as defined in 6.18) gives a holomorphic branch g_k of $[z^\alpha]$, viz.

$$g_k(z) = e^{\alpha[\log r + i(\theta + 2k\pi)]}(0 \neq z = re^{i\theta}, -\pi < \theta \leq \pi).$$

When α is rational there are only finitely many different branches, otherwise there are infinitely many.

(2) **The nth root** The special case in (1) when $\alpha = 1/n$, where $n = \pm 2, \pm 3, \ldots$, is of sufficient importance to warrant a closer analysis. We have $[z^{1/n}] = \{|z|^{1/n}e^{i\theta/n} : \theta \in [\arg z]\}$; there are n values for each $z \neq 0$. We define a family of multibranches by

$$F_k(r, \theta) = e^{2k\pi i/n} r^{1/n} e^{i\theta/n} \quad (k = 0, 1, \ldots, n-1).$$

These functions F_k have the properties stipulated in 6.19. If $\theta \in [\arg z]$ is chosen to vary continuously with z as z traces out

$\gamma(0; r)^*$, θ increases by 2π. We have

$$F_k(r,\ \theta+2\pi)=F_{k+1}(r,\ \theta)\quad\text{for}\quad 0\le k<n-1\quad\text{and}$$
$$F_{n-1}(r,\ \theta+2\pi)=F_0(r,\ \theta).$$

Hence $\gamma=\gamma(0;r)$ induces a cyclic permutation of the multibranches:

$$F_0 \underset{\gamma}{\to} F_1,\ F_1 \underset{\gamma}{\to} F_2,\ \ldots,\ F_{n-1}\underset{\gamma}{\to} F_0.$$

We conclude that 0 is a branch point. Replacing z by $1/z$ we see that ∞ is also a branch point. Cut the plane along $(-\infty, 0]$. An associated set of holomorphic branches is given, for $k = 0, 1, \ldots, n-1$, by

$$g_k(z)=e^{2k\pi i/n}r^{1/n}e^{i\theta/n}\qquad(0\ne z=re^{i\theta},\ -\pi<\theta\le\pi).$$

(3) Let w be the multifunction given by $[w(z)]=[[(z-a)(z-b)]^{\frac{1}{2}}]$ $(a,\ b\in\mathbb{C},\ a\ne b)$. Here $S=\mathbb{C}\backslash\{a,\ b\}$. Then for $z\in S$

$$[w(z)]=\{|(z-a)(z-b)|^{\frac{1}{2}}e^{\frac{1}{2}i(\theta+\phi)}:\theta\in[\arg(z-a)]\text{ and}$$
$$\phi\in[\arg(z-b)]\}.$$

We take our multibranches to be F_1 and F_2, where

$$F_1(r,\ R,\ \theta,\ \phi)=(rR)^{\frac{1}{2}}e^{\frac{1}{2}i(\theta+\phi)}\quad\text{and}\quad F_2(r,\ R,\ \theta,\ \phi)=-(rR)^{\frac{1}{2}}e^{\frac{1}{2}i(\theta+\phi)};$$

these functions have the properties stipulated in 6.19. To locate the branch points and suitable positions for cuts, we take contours γ_1, γ_2, γ_3, and γ_4 as shown in Fig. 6.4. The effect of moving z

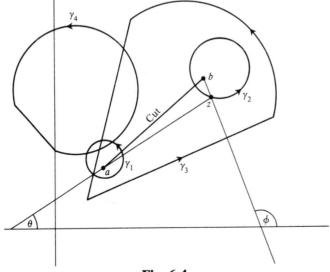

Fig. 6.4

round the image of each of these, with $\theta \in [\arg(z-a)]$ and $\phi \in [\arg(z-b)]$ varying continuously with z, is tabulated below.

	γ_1	γ_2	γ_3	γ_4
$\theta\uparrow$	2π	0	2π	0
$\phi\uparrow$	0	2π	2π	0
$\frac{1}{2}(\theta+\phi)\uparrow$	π	π	2π	0
	$F_1 \underset{\gamma_1}{\leftrightarrow} F_2$	$F_1 \underset{\gamma_2}{\leftrightarrow} F_2$	unchanged	unchanged

We deduce that a and b are branch points, and that contours γ_1 and γ_2 are 'bad', while γ_3 and γ_4 are not. Further $[\tilde{w}(z)] = [((1-az)(1-bz))^{\frac{1}{2}}/z]$, which has its branch points only at $1/a$ and $1/b$, so ∞ is not a branch point for w. We outlaw the 'bad' contours by cutting the plane between the two branch points, along $[a, b]$. In the cut plane w has two holomorphic branches, obtained from F_1 and F_2 by suitably restricting θ and ϕ. If, for example we have $a = 1$, $b = -1$, so our cut is along $[-1, 1]$, then we would take holomorphic branches f_1 and f_2 given by $f_1(z) = +|z^2 - 1|^{\frac{1}{2}}e^{\frac{1}{2}i(\theta+\phi)}$ and $f_2(z) = -|z^2 - 1|^{\frac{1}{2}}e^{\frac{1}{2}i(\theta+\phi)}$, where $z - 1 = |z - 1|e^{i\theta}$ ($-\pi < \theta \leq \pi$) and $z + 1 = |z + 1|e^{i\phi}$ ($0 \leq \phi < 2\pi$); here we have used 'upper edge values' on the cut. See Fig. 6.5(a).

(4) Let w be given by $[w(z)] = [\log(z^2 - 1)]$. Here $S = \mathbb{C}\backslash\{1, -1\}$. We have an infinite family of multibranches given by

$$F_k(r, R, \theta, \phi) = \log rR + i(\theta + \phi + 2k\pi) \qquad (k \in \mathbb{Z}).$$

We leave it to the reader to verify that the 'bad' contours are those which encircle one or both $+1$ or -1. Thus $+1$ and -1 are branch points. Also $[\tilde{w}(z)] = [\log[(1 - z^2)/z^2]]$, and this does have a branch point at 0, so ∞ is a branch point for w. We therefore take

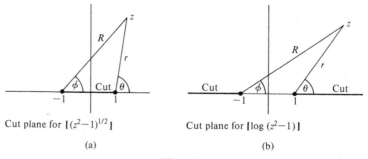

Cut plane for $[(z^2-1)^{1/2}]$

(a)

Cut plane for $[\log(z^2-1)]$

(b)

Fig. 6.5

cuts along $(-\infty, -1]$ and along $[1, \infty)$. (Compare this with (3), where ∞ was not a branch point.) The holomorphic branches are

$$f_k(z) = \log |z^2 - 1| + i(\theta + \phi + 2k\pi), \qquad (z \neq \pm 1, \, k \in \mathbb{Z})$$

where $z - 1 = |z - 1|e^{i\theta}$ $(0 \leqslant \theta < 2\pi)$ and $z + 1 = |z + 1|e^{i\phi}$ $(-\pi < \phi \leqslant \pi)$. See Fig. 6.5(b).

6.21 Remarks

In summary, given a multifunction w, we have sought to:
  (i) define a set of multibranches,
  (ii) locate the branch points,
  (iii) cut the plane so as to forbid 'bad' contours, and
  (iv) specify a holomorphic branch.
Here (i)–(iii) are stepping stones to our goal, (iv), and in simple cases not all of them may be needed. Multibranches are introduced merely to help us to see how to impose argument restrictions to obtain a continuous 1-valued selection of values from w. It may be possible immediately to specify a branch in terms of argument restrictions; cutting the plane then merely serves as a reminder of these constraints.

We have been somewhat schizophrenic about the status of points on cuts. To define branches on cuts, albeit in a non-unique way, as we did earlier, is theoretically tidy. However, for certain applications, it makes more sense to give a branch *two* values at each point of a cut (other than an endpoint), one for each edge. Consider the following example. In the plane cut along $[0, \infty)$ we take f to be the branch of the logarithm given by

$$f(z) = \log r + i\theta \qquad (0 \neq z = re^{i\theta}, 0 \leqslant \theta < 2\pi).$$

We cannot integrate f round $\gamma(0; R)$, since this crosses the cut. Suppose instead we integrate f round the contour shown in Fig. 6.6. In the cut plane we can take the horizontal lines to be at an arbitrarily small distance $\delta > 0$ from the real axis. We have

$$\lim_{\delta \to 0} \int_{AB} f(z) \, dz = \int_\varepsilon^R \lim_{\theta \downarrow 0} f(re^{i\theta}) \, dr = \int_\varepsilon^R \log r \, dr$$

and

$$\lim_{\delta \to 0} \int_{CD} f(z) \, dz = \int_\varepsilon^R \lim_{\theta \uparrow 2\pi} f(re^{i\theta}) \, dr = \int_\varepsilon^R (\log r + 2\pi i) \, dr.$$

Rather than go through this sort of limiting process every time we

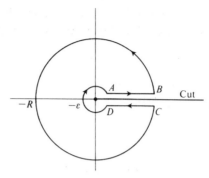

Fig. 6.6

integrate in a cut plane—and we do so quite frequently in Chapters 8 and 9—we allow ourselves the liberty of integrating 'along the edges' of a cut, using the obvious edge values in each case.

In this section we have treated multifunctions involving logarithms directly in terms of their sets of values. It is instructive to view such multifunctions also in terms of indefinite integrals. The treatment of the logarithm in 4.15 adopts this approach. Consider, more generally, $[w(z)]$ given by the possible values of $\int_{\gamma(z)} R(\zeta)\,d\zeta$, where $\gamma(z)$ is a path from some fixed point a to z and R is meromorphic or is a branch of a multifunction. Here many-valuedness arises because different paths $\gamma(z)$ can give different values of w. Take, as an example, the multifunction defining the inverse tangent: $[\int_{\gamma(z)} (1+\zeta^2)^{-1}\,d\zeta : \gamma(z)$ joins 1 and $z]$. Two paths from 1 to z give different values if one of i or $-$i lies between them. Cuts in the ζ-plane along half-lines from $\pm$i to ∞ prevent this, and an infinite family of branches of $\tan^{-1}$ can be defined in the cut plane.

6.22 An alternative perspective

It could be argued that we have attacked the many-valuedness problem in an ostrich-like way. Maybe, instead of working with an individual branch of a multifunction in a cut plane, it would be better to keep all branches in play, with one copy of the plane on which to define each. In other words, we treat the aggregate of branches as a single function on a domain set consisting of many copies of the plane. These copies are glued together so that in moving from one to another, we pass continuously from one branch of the multifunction to another. The resulting structure is known as a Riemann surface for the multifunction. A multi-storey

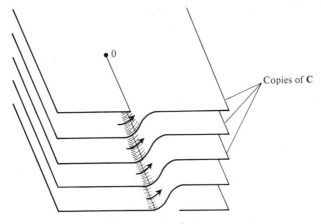

Riemann surface for log z

Fig. 6.7

car park provides a good mental picture. The floors of the car park represent copies of the plane, and the ramps taking one up and down between levels indicate how these copies are pasted together. The Riemann surface for the logarithm is modelled by a car park with infinitely many floors each of infinite extent, with a ramp joining each floor to the next; see Fig. 6.7. For more complicated multifunctions the car park designer might be said to have a warped sense of humour. An extensive theory of Riemann surfaces exists and provides a framework for an advanced treatment of multifunctions affording, in return for some effort, far greater insights than the naive plane-cutting approach. However, for the sort of multifunction problems discussed in this book, cut planes serve quite adequately.

Exercises

1. Find the Laurent expansion of $(z^2-1)^{-2}$ valid (i) for $0<|z-1|<2$, (ii) for $|z+1|>2$.

2. Find the principal part of the Laurent expansion about the indicated point a of
(i) $1/(z^2 \sin z)$ $(a=0)$, (ii) $(e^z-1)^{-2}$ $(a=0)$, (iii) e^{-1/z^4} $(a=0)$,
(iv) $\dfrac{e^z-1}{e^z+1}$ $(a=\pi i)$, (v) $\dfrac{e^{iz}}{(z^2+b^2)^3}$ $(a=ib)$.

3. By considering $e^{z-1/z}$ or otherwise, prove that, for $n=1, 2, \ldots,$

$$\frac{1}{2\pi} \int_0^{2\pi} \cos(n\theta - 2\sin\theta)\, d\theta = \sum_{k=0}^{\infty} \frac{(-1)^k}{k!(n+k)!}.$$

4. Determine the orders of the zeros of the functions in Exercise 2.11.

5. Locate and classify the singularities in $\mathbb{C}$ of the following functions:

(i) $\dfrac{1}{z^3(z^2 + 1)}$, (ii) $\dfrac{1 - e^{iz}}{z^2}$, (iii) $\dfrac{e^{iz}}{(z^2 + z + 1)^2}$,

(iv) $\dfrac{\cot \pi z}{z - 1}$, (v) $\dfrac{z \sin z}{\cos z - 1}$, (iv) $\dfrac{1}{1 - e^{z^2}}$.

6. Locate and classify the singularities, including singularities at ∞, of the following functions:

(i) $\dfrac{1}{z^4 - z^2 + 1}$, (ii) $\dfrac{e^z}{1 - z}$, (iii) $\tan^2 z$, (iv) $\cosh \dfrac{1}{z}$,

(v) $\dfrac{1}{(\pi + z)\sin z} - \dfrac{1}{\pi z}$, (vi) $\dfrac{z^2}{\cosh z - \cos z}$, (vii) $e^{\cosec 1/z}$.

7. Construct functions f_1, f_2, f_3, and f_4 such that
(i) f_1 is holomorphic in $\mathbb{C}$ except for a pole of order 4 at each point $2k + 1$ ($k \in \mathbb{Z}$) and a removable singularity at 0,
(ii) f_2 has simple poles at ± 1, $\pm i$ and a double pole at ∞, but is otherwise holomorphic,
(iii) f_3 is holomorphic apart from isolated essential singularities at 0, 1, ∞,
(iv) f_4 is holomorphic in $\mathbb{C}$ except for non-isolated essential singularities at ± 1 and a set of double poles.

8. Suppose that, for $R < |z| < S$, $f(z) = g(z) + h(z)$, where $g(z)$ is holomorphic for $|z| < S$ and $h(z)$ is holomorphic and bounded for $|z| > R$. Let $\langle c_n \rangle$ be the Laurent coefficients of f in the annulus $\{z : R < |z| < S\}$. Prove that, for some constant c,

$$g(z) = c + \sum_{n=1}^{\infty} c_n z^n \qquad (|z| < S).$$

9. Suppose that f is continuous and bounded in $D(a; r)$ and that f is holomorphic in $D'(a; r)$. By considering the Laurent expansion of f about a, prove that f is holomorphic in $D(a; r)$.

10. Suppose that f is holomorphic in a punctured disc, centre a. Let $w \in \mathbb{C}$ be given. Suppose that there exist $\varepsilon > 0$ and $r > 0$ such that $|f(z) - w| \geq \varepsilon$ for all $z \in D'(a; r)$. By considering the function $(f - w)^{-1}$, prove that f cannot have an essential singularity at a. Deduce the Casorati–Weierstrass theorem stated in 6.12(3).

11. Let γ be the circular path $\gamma(0; 2)$. Suppose values $f(z) \in [w(z)]$ are selected so that $f(\gamma(t))$ varies continuously as t increases from 0 to 2π, with $f(\gamma(0))$ real. Determine the initial value, $f(\gamma(0))$, and final value, $f(\gamma(2\pi))$, in case $[w(z)]$ is (i) $[(z - 1)^{1/3}]$, (ii) $[\log(z^{-1})]$, (iii) $[[z(z^2 - 1)]^{1/2}]$, (iv) $[z^{\sqrt{2}}]$. (In (iii), assume further that $f(\gamma(0))$ is positive.)

12. Verify that the multifunctions below have branch points as indicated and that the cuts suggested forbid 'bad' contours. In each case specify a holomorphic branch.

(i) $[(z - i)^{1/2}]$ (branch points i; ∞, cut along $\{iy : y \geqslant 1\}$).

(ii) $[[(z - 1)/(z + 1)]^{1/4}]$ (branch points ± 1; cut along $[-1, 1]$).

(iii) $[[z(z - 1)]^{-1/2}]$ (branch points 0, 1; cut along $[0, 1]$).

(iv) $[[z(z - 1)]^{2/3}]$ (branch points $0, 1, \infty$; cuts along $(-\infty, -1]$ and $[1, \infty)$).

(v) $[\log[(1 + iz)/(1 - iz)]]$ (branch points ± 1; cuts along $\{iy : y$ real, $|y| \geqslant 1\}$).

13. For each of the following multifunctions, locate the branch points, suggest how the plane should be cut, and specify a holomorphic branch:
(i) $[(z^2 - 1)^{-1/2}]$, (ii) $[[(z - 1)/(z + 1)]^{\pi}]$, (iii) $[\log(1 + z^2)]$.

14. Obtain a power series expansion, valid in $D(0; 1)$, of some holomomorphic branch of $[\log(1 + z)]$.

7 Cauchy's residue theorem

The chief result in this chapter is Cauchy's residue theorem which does for functions with poles what Cauchy's theorem does for holomorphic functions. This theorem is extremely important: theoretical and practical applications of it occupy much of the rest of the book.

To use Cauchy's residue theorem effectively one needs to be able to calculate residues (as defined in 7.3) with a minimum of effort, and to manipulate complex integrals efficiently. The tricks of the trade are set out in the last two sections, ready for use in Chapters 8 and 9. To provide some motivation, we preface this chapter with a simple example previewing Chapter 8.

Cauchy's residue theorem

7.1 Example

Suppose we wish to evaluate

$$I := \int_0^\infty (1+x^4)^{-1} \, dx = \lim_{R \to \infty} \int_0^R (1+x^4)^{-1} \, dx.$$

(In Riemann integration the limit on the right-hand side defines the integral, in Lebesgue integration the above equation is a consequence of the Monotone convergence theorem.) Let γ be the semicircular contour shown in Fig. 7.1. We have

$$\int_\gamma (1+z^4)^{-1} = \int_{-R}^R (1+x^4)^{-1} \, dx + \int_0^\pi (1+R^4 e^{4i\theta})^{-1} Rie^{i\theta} d\theta.$$

As $R \to \infty$, the first term on the right-hand side converges to $2I$, while the second, as we showed in 3.11(1), tends to zero. We should therefore like to evaluate $\int_\gamma (1+z^4)^{-1} \, dz$, which is a contour integral round two poles of the integrand, at the zeros of $(1+z^4)$ in the upper half-plane (by 6.10), that is, at $e^{\pi i/4}$ and $e^{3\pi i/4}$. Cauchy's residue theorem, 7.4, is just what we need. It gives a formula for $\int_\gamma f(z) \, dz$ when f is holomorphic except for a finite number of

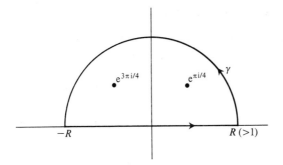

Fig. 7.1

poles inside γ. (The computation of I is completed in 7.12(2), after we have proved the Residue theorem and developed techniques for applying it.)

7.2 Lemma

Let f be holomorphic inside and on a positively oriented contour γ except at the point a inside γ, where it has a pole, and let

$$f(z) = \sum_{n=-m}^{\infty} c_n(z-a)^n$$

be the Laurent expansion of f about a. Then

$$\int_{\gamma} f(z)\,dz = 2\pi i c_{-1}.$$

Proof. Choose r such that $\bar{D}(a;r) \subseteq I(\gamma)$. Then

$$\int_{\gamma} f(z)\,dz = \int_{\gamma(a;r)} f(z)\,dz \quad \text{(by the Deformation theorem (I), 4.8)}$$

$$= \int_{\gamma(a;r)} \sum_{n=-m}^{\infty} c_n(z-a)^n\,dz$$

$$= \sum_{n=-m}^{\infty} c_n \int_{\gamma(a;r)} (z-a)^n\,dz \quad \text{(by 3.13)}$$

$$= 2\pi i c_{-1} \quad \text{(by 3.7(1)).} \qquad \square$$

7.3 Definition

Suppose $f \in H(D'(a;r))$ has a pole at a. The *residue* of f at a is the (unique) coefficient c_{-1} of $(z-a)^{-1}$ in the Laurent expansion of f about a, and is denoted $\text{res}\{f(z); a\}$.

7.4 Cauchy's residue theorem

Let f be holomorphic inside and on a positively oriented contour γ except for a finite number of poles at $a_1, \ldots, a_m$ inside γ. Then

$$\int_\gamma f(z)\,dz = 2\pi i \sum_{k=1}^{m} \text{res}\{f(z); a_k\}.$$

Proof. Let $f_k(z)$ be the principal part of the Laurent expansion of f about a_k. Then

$$g := f - \sum_{k=1}^{m} f_k$$

has only removable singularities at $a_1, \ldots, a_m$; remove them (see 6.6 and 6.12(1)). By Cauchy's theorem, $\int_\gamma g(z)\,dz = 0$. Hence

$$\int_\gamma f(z)\,dz = \sum_{k=1}^{m} \int_\gamma f_k(z)\,dz = 2\pi i \sum_{k=1}^{m} \text{res}\{f(z); a_k\},$$

by Lemma 7.2, applied to each f_k. $\square$

7.5 Example

To find those functions satisfying (i) f is holomorphic in $\mathbb{C}$ except for simple poles at the cube roots of unity, viz. 1, $\omega = e^{2\pi i/3}$, and $\omega^2 = e^{4\pi i/3}$, where it has residues 1, α, and β, respectively, where $\alpha\beta = 1$, and (ii) there exists a constant K such that $|z^2 f(z)| \le K$ for $|z| \ge 2$.

Solution. By (i), Cauchy's residue theorem, and 3.9, for $R \ge 2$,

$$|2\pi i(1 + \alpha + \alpha^{-1})| = \left| \int_{\gamma(0;R)} f(z)\,dz \right| \le \int_0^{2\pi} |f(Re^{i\theta})|\, R\, d\theta$$

$$\le 2\pi K/R \quad \text{(by (ii)).}$$

This forces $1 + \alpha + \alpha^{-1} = 0$, whence $\alpha = \omega$ or $\alpha = \omega^2$, and $\beta = \alpha^2$.

Following the proof of Theorem 6.16, we let

$$g(z) := f(z) - \frac{1}{z-1} - \frac{\alpha}{z-\omega} - \frac{\alpha^2}{z-\omega^2}.$$

Then g has only removable singularities, and can be regarded as holomorphic in $\mathbb{C}$. Condition (ii) implies that $g(z) \to 0$ as $|z| \to \infty$. Combining this with 1.18, we see that g is bounded in $\mathbb{C}$. By Liouville's theorem, 5.2, g is a constant, and so must be identically zero. Put in the values for α found above and simplify. The possible forms for f turn out to be

$$f(z) = 3(z^3 - 1)^{-1} \quad \text{and} \quad f(z) = 3z(z^3 - 1)^{-1}.$$

Clearly each of these functions does satisfy (i) and (ii). □

Counting zeros and poles

The theorems in this section have close affinities with the results on argument in Chapter 4. Exercise 7.13 is provided for those interested in pursuing these connections. Here we adopt a less sophisticated approach.

7.6 Theorem

Let f be holomorphic inside and on a positively oriented contour γ except for P poles inside γ. Let f be non-zero on γ and have N zeros inside γ. Then

$$\frac{1}{2\pi i} \int_\gamma \frac{f'(z)}{f(z)} \, dz = N - P.$$

(A pole or zero of order m is counted m times.)

Proof. The function f'/f is holomorphic inside and on γ, except at the poles and zeros of f lying inside γ.

If a is a zero of f of order m, then there exists a function g which is holomorphic and non-zero in some $D(a; r)$ and such that

$$f(z) = (z - a)^m g(z) \quad \text{in } D(a; r).$$

Then

$$\frac{f'(z)}{f(z)} = \frac{m}{z - a} + \frac{g'(z)}{g(z)}.$$

Since $g'/g \in H(D'(a; r))$, we conclude that f'/f has a simple pole at a of residue m. Similarly, if b is a pole of f of order n, f'/f has a simple pole at b of residue $-n$.

The theorem now follows from Cauchy's residue theorem, 7.4. □

7.7 Rouché's theorem

Let f and g be holomorphic inside and on a contour γ and suppose $|f(z)| > |g(z)|$ on γ^*. Then f and $f + g$ have the same number of zeros inside γ. (That the number of zeros is finite is a consequence of the Bolzano–Weierstrass theorem, 1.17, and the Identity theorem, 5.14, applied with some care.)

Proof. Let $t \in [0, 1]$. Since $|f(z)| > |g(z)|$ on γ^*, $(f + tg)(z) \neq 0$ for any $z \in \gamma^*$. Assume without loss of generality that γ is positively oriented, and define

$$\phi(t) = \frac{1}{2\pi i} \int_\gamma \frac{(f' + tg')(z)}{(f + tg)(z)} \, dz.$$

By 7.6, $\phi(t)$ is the number of zeros of $f + tg$ inside γ. The function ϕ is integer-valued, and, if it is continuous, it must be constant (see 1.19). In this event, the number of zeros of f inside γ equals $\phi(0) = \phi(1)$ which is the number of zeros of $f + g$ inside γ, as required.

It is possible to establish continuity of ϕ by citing a general theorem about functions defined by integrals. Here, alternatively, is a direct proof. Fix t and consider

$$\phi(t) - \phi(s) = \frac{t - s}{2\pi i} \int_\gamma \frac{(g'f - f'g)(z)}{(f + tg)(z)(f + sg)(z)} \, dz.$$

By 1.18, we can find positive constants M and m such that for all $z \in \gamma^*$, $|(g'f - f'g)(z)| \leq M$, $|g(z)| \leq M$, and $|(f + tg)(z)| \geq m$. Then

$$|(f + sg)(z)| \geq |(f + tg)(z)| - |s - t| \, |g(z)| \quad \text{(by 1.4(3))}$$
$$\geq \tfrac{1}{2}m \quad \text{if } |s - t| \leq \tfrac{1}{2}m/M.$$

Hence, for $|s - t|$ sufficiently small,

$$|\phi(t) - \phi(s)| \leq \frac{|t - s| \, M}{\pi m} \times \text{length}(\gamma) \quad \text{(by 3.10)},$$

whence ϕ is continuous at t. $\qquad\square$

Rouché's theorem is needed in Chapter 10 in the proofs of some important theorems. The example below indicates the way in which it can be used to locate zeros of particular functions.

7.8 Example

To show that $2 + z^2 - e^{iz}$ has precisely one zero in the open upper half-plane.

Solution. Take $f(z) = 2 + z^2$, $g(z) = -e^{iz}$, and γ the semicircular contour shown in Fig. 7.1. For $z \in [-R, R]$,

$$|f(z)| \geq 2 > 1 = |g(z)|,$$

and for $z = Re^{i\theta}$ $(0 \le \theta \le \pi)$,

$$|f(z)| \ge R^2 - 2 > 1 \ge e^{-R\sin\theta} = |g(z)|,$$

so long as $R > \sqrt{3}$. With this restriction in force, Rouché's theorem can be applied. We deduce that $f(z) + g(z) = 2 + z^2 - e^{iz}$ has the same number of zeros in $\{z : \mathrm{Im}\, z > 0, |z| < R\}$ $(R > \sqrt{3})$ as has $f(z) = 2 + z^2$, that is, just one. Hence $2 + z^2 - e^{iz}$ has precisely one zero in the open upper half-plane. □

Calculation of residues

It is inconvenient to have to resort to the Laurent expansion to find a residue, as anyone who tries to complete Example 7.1 by this method will quickly discover. In this section we derive formulae which enable residues to be calculated with a minimum of fuss.

7.9 Classification of poles

Assume f has a pole of order m at a. We call the pole:

   simple if $m = 1$, *multiple* otherwise;

   type I (overt) if $f(z)$ is expressed in the form $g(z)(z-a)^{-m}$ where $g \in H(D(a\,;\,r))$ for some r and $g(a) \ne 0$, and *type II* (covert) otherwise.

   If a is a type II pole of order m, then in some disc $D(a\,;\,r)$, $f(z)$ can be expressed as $h(z)/k(z)$, where h and k are in $H(D(a\,;\,r))$, $h(a) \ne 0$, and k has a zero of order m at a.

   Whether a pole of f is overt or covert is a matter of how $f(z)$ is written. Covert poles can often be converted to overt poles by factorization, but it is seldom prudent to do this in residue calculations (see Examples 7.12).

Examples

(1) $1/[(z-i)(z+i)]$ has simple poles, type I, at $\pm i$; $1/(z^2+1)$ has simple poles, type II, at $\pm i$.
(2) $\tan^2 z$ has double poles, type II, at $\frac{1}{2}(2n+1)\pi$ $(n \in \mathbb{Z})$.

7.10 The residue at a simple pole

Let $f \in H(D'(a\,;\,r))$ and have a simple pole at a.

Lemma

$$\text{res}\{f(z); a\} = \lim_{z \to a} (z - a)f(z).$$

Proof. In $D'(a; r)$, $f(z) = \sum_{n=-1}^{\infty} c_n(z - a)^n$. This implies

$$\lim_{z \to a} (z - a)f(z) = c_{-1} (:= \text{res}\{f(z); a\}).$$ □

(1) **Simple pole, type I**   If $f(z) = g(z)/(z - a)$, where $g \in H(D(a; r))$ and $g(a) \neq 0$, then

$$\text{res}\{f(z); a\} = g(a).$$

Proof. The formula is immediate from the lemma above.   □

(2) **Simple pole, type II**   If $f(z) = h(z)/k(z)$, where h and k are in $H(D(a; r))$, $h(a) \neq 0$, $k(a) = 0$ and $k'(a) \neq 0$, then

$$\text{res}\{f(z); a\} = \frac{h(a)}{k'(a)}.$$

Proof.

$$\text{res}\{f(z); a\} = \lim_{z \to a} (z - a)\frac{h(z)}{k(z)} \quad \text{(by the lemma)}$$

$$= \lim_{z \to a} h(z) \frac{z - a}{k(z) - k(a)}$$

$$= \frac{h(a)}{k'(a)}.$$ □

7.11   The residue at a multiple pole

Let f have a pole of order $m > 1$ at a.

(1) **Multiple pole, type I**   Let $f(z) = (z - a)^{-m}g(z)$, where $g \in H(D(a; r))$ and $g(a) \neq 0$. Then

$$\text{res}\{f(z); a\} = \frac{1}{(m - 1)!} g^{(m-1)}(a).$$

Proof. By Cauchy's formula for derivatives, 5.4,

$$g^{(m-1)}(a) = \frac{(m - 1)!}{2\pi i} \int_{\gamma(a;\frac{1}{2}r)} \frac{g(z)}{(z - a)^m} dz$$

$$= \frac{(m - 1)!}{2\pi i} \int_{\gamma(a;\frac{1}{2}r)} f(z) dz$$

$$= (m - 1)!\,\text{res}\{f(z); a\} \quad \text{(by Lemma 7.2).}$$ □

(2) **Multiple pole, type II**   No formula as neat as that for a type II simple pole exists. To find the residue, either convert to a type I pole or compute c_{-1} in the Laurent expansion.

7.12   Examples

(1) $f(z) = 1/[(2-z)(z^2+4)]$ has simple poles at 2 (type I) and at $\pm 2i$ (type II).

$$\text{res}\{f(z); 2\} = -\frac{1}{2^2+4} = -\tfrac{1}{8} \quad \text{(by 7.10(1))}$$

and

$$\text{res}\{f(z); \pm 2i\} = \frac{1}{(2 \mp 2i)(\pm 4i)} = \frac{1 \mp i}{16} \quad \text{(by 7.10(2))}.$$

(2) $f(z) = 1/(1+z^4)$ has simple poles, type II, at the points $z_k = e^{(2k+1)\pi i/4}$ $(k = 0, 1, 2, 3)$. By 7.10(2),

$$\text{res}\{f(z); z_k\} = \left[\frac{1}{4z^3}\right]_{z=z_k} = -\tfrac{1}{4}e^{(2k+1)\pi i/4}, \quad \text{since } z_k^4 = -1.$$

Factorization to convert the poles to type I is not recommended here. Note that we can now complete Example 7.1: by Cauchy's residue theorem,

$$\int_\gamma \frac{1}{1+z^4}\, dz = 2\pi i\{\text{res}\{f(z); z_1\} + \text{res}\{f(z); z_2\}\}$$

$$= -\tfrac{1}{2}\pi i(e^{\pi i/4} + e^{3\pi i/4}).$$

Hence, from 7.1,

$$2\int_0^\infty \frac{1}{1+x^4}\, dx = \frac{\pi}{\sqrt{2}}.$$

(3) $f(z) = e^{iz}z^{-4}$ has a type I pole of order 4 at 0. By 7.11(1),

$$\text{res}\{f(z); 0)\} = \frac{1}{3!}\left[\frac{d^3}{dz^3} e^{iz}\right]_{z=0} = -\frac{i}{6}.$$

Alternatively, the Laurent expansion

$$\frac{e^{iz}}{z^4} = \frac{1}{z^4} + \frac{i}{z^3} - \frac{1}{2!z^2} - \frac{i}{3!z} + \dots \quad (0 < |z|)$$

gives

$$\text{res}\{f(z); 0\} = -\frac{i}{6} \quad \text{(by 7.3)}.$$

(4) $f(z) = (z+1)^{-2}(z^3-1)^{-1}$ has a double pole, type I, at -1 and simple poles, type II, at 1, ω, and ω^2, where $\omega^3 = 1$. By 7.11(1),

$$\operatorname{res}\{f(z); -1\} = \left[\frac{d}{dz}\left(\frac{1}{z^3-1}\right)\right]_{z=-1} = \left[\frac{-3z^2}{(z^3-1)^2}\right]_{z=-1} = -\tfrac{3}{4}.$$

Suppose α is any one of the cube roots of unity: 1, ω, or ω^2. Then

$$\operatorname{res}\{f(z); \alpha\} = \frac{1/(1+\alpha)^2}{3\alpha^2} = \frac{1}{3(\alpha^2+2+\alpha)} \quad \text{(by 7.10(2))}.$$

So $\operatorname{res}\{f(z); 1\} = \tfrac{1}{12}$ and $\operatorname{res}\{f(z); \omega\} = \operatorname{res}\{f(z); \omega^2\} = \tfrac{1}{3}$ (remember $\alpha^3 = 1$ ($\alpha \neq 1$) implies $1 + \alpha + \alpha^2 = 0$).

(5) $f(z) = (\pi \cot \pi z)/z^2$ has a triple pole, type II, at 0 and simple poles, type II, at $n = \pm 1, \pm 2, \ldots$. By 7.10(2),

$$\operatorname{res}\{f(z); n\} = \left[\pi \frac{(\cos \pi z)/z^2}{\pi \cos \pi z}\right]_{z=n} = \frac{1}{n^2} \quad (n \in \mathbb{Z}, n \neq 0).$$

Near $z = 0$,

$$\frac{\pi \cot \pi z}{z^2} = \frac{1}{z^3} - \frac{\pi^2}{3z} + \ldots \quad \text{(by 6.4(4))}.$$

Hence

$$\operatorname{res}\{f(z); 0\} = -\pi^2/3.$$

7.13 Remark

Cauchy's residue theorem provides the natural way to attack a contour integral $\int_\gamma f(z)\,dz$ when f has either several poles, or a single type II pole, inside γ. For an integral round just one type I pole Cauchy's integral formula, 5.1, or Cauchy's formula for derivatives, 5.4, should be used; it is devious to cite the Residue theorem in this situation (see the proof of 7.11(1)). Of course, when f has no poles inside or on γ, Cauchy's theorem applies.

Estimation of integrals

In Residue theorem applications we frequently need to find the limiting value of an integral along a path as that path shrinks or expands indefinitely. This section surveys the standard ways of manipulating integrals in these circumstances.

7.14 Basic inequalities

(1) The following are true for any complex numbers $z_1, z_2, \ldots$.
- (i) $|z_1 + z_2| \leq |z_1| + |z_2|$ (1.4(2)).
- (ii) $|z_1 + \ldots + z_n| \leq |z_1| + \ldots + |z_n|$ (by induction on (i)).
- (iii) $|z_1 + z_2| \geq ||z_1| - |z_2||$ (1.4(3)).
- (iv) $|z_1 + \ldots + z_n| \geq |z_1| - |z_2| - \ldots - |z_n|$ (by (ii) and (iii)).
- (v) $|z_1| \leq |z_2| \Leftrightarrow 1/|z_1| \geq 1/|z_2|$ $\quad (z_1, z_2 \neq 0)$.

Inequalities (iii) and (iv) (often used in conjunction with (v)) are extremely useful but frequently misquoted; note the minus signs. If in doubt when handling, for example $1/|z_1 + z_2|$, remember that to make this bigger one must make the denominator smaller, and that this is not in general achieved by replacing $|z_1 + z_2|$ by $|z_1| + |z_2|$. This is an opportune time to issue a reminder that inequalities must be between real numbers; see 1.4. Omission of moduli from (i)–(v) causes havoc.

(2) Suppose γ is a path with parameter interval $[\alpha, \beta]$ and f is a continuous function on γ^*. Then

$$\left| \int_\gamma f(z)\, dz \right| \leq \int_\alpha^\beta |f[\gamma(t)]\gamma'(t)|\, dt \quad \text{(Estimation theorem, 3.9)}$$

$$\leq \sup_{z \in \gamma^*} |f(z)| \times \text{length}(\gamma) \quad \text{(Corollary 3.10)}.$$

(3) **Jordan's inequality** Suppose $0 < \theta \leq \frac{1}{2}\pi$. Then

$$\frac{2}{\pi} \leq \frac{\sin \theta}{\theta} \leq 1.$$

Proof. It will be sufficient to prove that, on $(0, \frac{1}{2}\pi]$, $(\sin \theta)/\theta$ decreases as θ increases. This is the case if

$$\frac{d}{d\theta}\left(\frac{\sin \theta}{\theta} \right) \leq 0 \quad \text{on} \quad (0, \frac{1}{2}\pi].$$

But

$$\frac{d}{d\theta}\left(\frac{\sin \theta}{\theta} \right) = \frac{\theta \cos \theta - \sin \theta}{\theta^2} \leq 0$$

whenever $\theta \cos \theta \leq \sin \theta$. Since $[\theta \cos \theta - \sin \theta]_{\theta=0} = 0$, it is now enough to note that, on $(0, \frac{1}{2}\pi]$, $\theta \cos \theta - \sin \theta$ has a non-positive derivative and so decreases as θ increases. □

7.15 Basic limits

The following limits are probably familiar to the reader. Since we shall use them repeatedly, we include proofs.
(1) For any positive constant k, $x^k e^{-x} \to 0$ as $x \to \infty$ $(x \in \mathbb{R})$.

(2) For any positive constant k,

$$x^{-k} \log x \to 0 \quad \text{as} \quad x \to \infty \quad \text{and} \quad x^k \log x \to 0 \quad \text{as} \quad x \to 0 \, (x \in \mathbb{R}).$$

Proof. (1) For $x > 0$, $0 < x^k e^{-x} = x^k/(1 + x + \ldots + x^n/n! + \ldots) < n! x^{k-n}$. This is true for any n. Since n can be chosen greater than k, the result follows.
(2) To obtain the first result, put $x = e^{y/k}$ and note that

$$x^{-k} \log x = (y e^{-y}/k),$$

which tends to zero as $y \to \infty$ and hence as $x \to \infty$. The second limit follows from the first if we replace x by x^{-1}. □

7.16 Estimation of integrals round large circular arcs

We have already encountered several integrals of this type. We now consider their behaviour more generally. Let f be continuous on γ^*, where $\gamma(\theta) = Re^{i\theta}$ $(\theta \in [\theta_1, \theta_2])$. We have $\gamma'(\theta) = iRe^{i\theta}$, so by 7.14(2),

$$\left| \int_\gamma f(z) \, dz \right| \leq \int_{\theta_1}^{\theta_2} |f(Re^{i\theta})| \, R \, d\theta. \tag{*}$$

Examples (1) Let $f(z) = (z^2 + z + 1)^{-2}$. By (*) and 7.14(1)(iv),

$$\left| \int_\gamma f(z) \, dz \right| \leq \int_{\theta_1}^{\theta_2} \frac{R}{(R^2 - R - 1)^2} \, d\theta \quad \text{(if } R^2 - R - 1 > 0\text{)}$$
$$= O(R^{-3}) \quad \text{for large } R.$$

(2) Let $f(z) = e^{iz} z^{-k}$ $(k > 0)$, $\theta_1 = 0$ and $\theta_2 = \pi$. Then

$$|f(Re^{i\theta})| = \left| \frac{e^{i(R \cos \theta + iR \sin \theta)}}{R^k e^{ik\theta}} \right| = \frac{e^{-R \sin \theta}}{R^k}.$$

On $[0, \pi]$, $\sin \theta \geq 0$, so $e^{-R \sin \theta} \leq e^0 = 1$. Hence

$$R \, |f(Re^{i\theta})| \leq R^{1-k}.$$

If $k > 1$, this estimate and (*) show that $\int_\gamma f(z)\,dz \to 0$ as $R \to \infty$. If $0 < k \leqslant 1$, the bound $e^{-R \sin \theta} \leqslant 1$ is too crude. However, Jordan's inequality can be used to give

$$\left| \int_\gamma f(z)\,dz \right| \leqslant \int_0^\pi e^{-R \sin \theta} R^{1-k}\,d\theta = 2 \int_0^{\frac{1}{2}\pi} e^{-R \sin \theta} R^{1-k}\,d\theta$$

$$\leqslant 2R^{1-k} \int_0^{\frac{1}{2}\pi} e^{-2R\theta/\pi}\,d\theta$$

$$\leqslant \pi R^{-k}(1 - e^{-R}),$$

and this tends to zero as $R \to \infty$.

Note that it was first necessary to change the range of integration to $[0, \frac{1}{2}\pi]$; Jordan's inequality is not valid on $[0, \pi]$.

One should invoke Jordan's inequality only when a rougher estimate is no help. In this example, it is not needed when $k > 1$, but, when $0 < k \leqslant 1$, it produces a life-saving factor R^{-1} (cf. 8.5, 8.9).

7.17 Estimation of integrals round small circular arcs

Cauchy's residue theorem assumes f to be holomorphic on γ. Thus when the function appropriate to our needs has a pole or branch point on the contour we should like to use, it is necessary to modify either the function or the contour. For example, to avoid a pole at a, we might make an indentation consisting of a small circular arc centre a; for an illustration see Example 8.5.

(1) **Indentation at a simple pole: Lemma** Let $f \in H(D'(a ; r))$ and let f have a simple pole of residue b at a. Let

$$\gamma_\varepsilon(\theta) := a + \varepsilon e^{i\theta} \quad (\theta \in [\theta_1, \theta_2]),$$

where $0 < \varepsilon < r$ and $0 \leqslant \theta_1 < \theta_2 \leqslant 2\pi$. Then

$$\lim_{\varepsilon \to 0} \int_{\gamma_\varepsilon} f(z)\,dz = ib(\theta_2 - \theta_1). \tag{†}$$

Proof. By 7.10, $b = \lim_{z \to a} (z - a)f(z)$. Let $g(z) = (z - a)f(z) - b$. Given $\eta > 0$, there exists $\delta > 0$ such that $|g(z)| < \eta$ whenever $0 < |z - a| < \delta$. Let $0 < \varepsilon < \min\{r, \delta\}$. When $z = \gamma_\varepsilon(\theta)$ we have $\gamma_\varepsilon'(\theta) = \varepsilon i e^{i\theta} = i(z - a)$, so

$$\left| \int_{\gamma_\varepsilon} f(z)\,dz - ib(\theta_2 - \theta_1) \right| = \left| \int_{\theta_1}^{\theta_2} \{f[\gamma_\varepsilon(\theta)]\gamma_\varepsilon'(\theta) - ib\}\,d\theta \right|$$

$$= \left| \int_{\theta_1}^{\theta_2} g[\gamma_\varepsilon(\theta)]\,d\theta \right|$$

$$< \eta(\theta_2 - \theta_1). \qquad \square$$

Note If $\gamma_\varepsilon = \gamma(a; \varepsilon)$, $\int_{\gamma_\varepsilon} f(z)\,dz = 2\pi i b$, by Cauchy's residue theorem. The lemma asserts that, more generally, the integral of f along γ_ε is, *in the limit as* $\varepsilon \to 0$, $2\pi i b$ times the fraction of $\gamma(a; \varepsilon)^*$ traversed.

(2) **Indentation at a multiple pole** The formula (†) in (1) above is *not* applicable in this case, as inspection of the proof shows. It is usually best to modify the function instead of indenting the contour.

(3) **Indentation at a branch point** Suppose f is a holomorphic branch of a multifunction with a branch point at a. Note that a is not an isolated singularity in the sense of 6.5, since $f \notin H(D'(a\,; r))$ for any r. Try basic inequalities, or (occasionally) adapt (1) above.

Exercises

1. Find the residues at the poles of the functions given in Exercise 6.5.
2. Compute

$$\text{(i)} \int_{\gamma(0;2)} \frac{1}{(z-1)^2(z^2+1)}\,dz, \qquad \text{(ii)} \int_{\gamma(0;8)} (1+e^z)^{-1}\,dz.$$

3. By considering the integral

$$\int_{\gamma(0;1)} \frac{z}{(2z^4+5z^2+2)}\,dz,$$

prove that

$$\int_0^{2\pi} \frac{1}{1+8\cos^2\theta}\,d\theta = \frac{2\pi}{3}.$$

4. Evaluate $\int_0^{2\pi} (\cos^4\theta + \sin^4\theta)\,d\theta$ by converting it into an integral round $\gamma(0; 1)$ and applying the Residue theorem.
5. Evaluate the following limits:

$$\text{(i)} \lim_{R\to\infty} \int_{\Gamma_R} \frac{e^{iz}}{(z^4+z^3+z^2+z+1)^2}\,dz, \qquad \text{(ii)} \lim_{R\to\infty} \int_{\gamma(0;R)} \frac{p(z)}{q(z)}\,dz$$

(where p and q are polynomials and $\deg p < \deg q - 1$),

$$\text{(iii)} \lim_{R\to\infty} \int_{\Gamma_R} \left(\frac{z-1}{z^2+1}\right) e^{iz}\,dz, \qquad \text{(iv)} \lim_{\varepsilon\to 0} \int_{\Gamma_\varepsilon} \operatorname{cosec}\pi z\,dz.$$

6. A function f is holomorphic in $\mathbb{C}$ except for double poles at 1 and -1 of residues a and b, respectively. It is also given that, for some constant

$K, |z^2 f(z)| \leq K$ for large $|z|$. Prove that $a + b = 0$. Find f if $a = 1$ and $f(2i) = f(-2i) = 0$.

7. Suppose p and q are polynomials of degrees m and n, respectively, where $n \geq m + 1$, and suppose q has simple zeros at $b_1, \ldots, b_n$. By integrating $f(w) = p(w)/(q(w)(w - z))$ round $\gamma(0; R)$ for large R, obtain the partial fraction decomposition.

$$\frac{p(z)}{q(z)} = \sum_{k=1}^{n} \frac{p(b_k)}{q'(b_k)} (z - b_k)^{-1}.$$

Hence decompose $\dfrac{1 - z^2}{1 + z^4}$ into partial fractions.

8. Let $f(z) = \operatorname{cosec} \pi z$ and let $\langle c_n \rangle$ and $\langle d_n \rangle$ be the Laurent coefficients of f in $\{z : 0 < |z| < 1\}$ and $\{z : 1 < |z| < 2\}$, respectively. Prove that, if n is odd, $\frac{1}{2}(c_n - d_n)\pi = 1$, and find the corresponding result when n is even.

9. Suppose f is holomorphic inside and on $\gamma(0; 1)$. By integration round the contour shown in Fig. 6.6 prove that

$$\int_0^1 f(x) \, dx = \frac{1}{2\pi i} \int_{\gamma(0; 1)} f(z)(\log z - i\pi) \, dz,$$

where $\log z$ denotes the branch of the logarithm whose imaginary part lies between 0 and 2π. Deduce that

$$\left| \int_0^1 f(x) \, dx \right| \leq \frac{1}{2} \int_0^{2\pi} |f(e^{i\theta})| \, d\theta.$$

10. Prove that the equation $z^5 + 15z + 1 = 0$ has precisely four solutions in the annulus $\{z : 3/2 < |z| < 2\}$.

11. Suppose $R > 0$ is given. Prove that, if N is sufficiently large,

$$\sum_{n=0}^{N} z^n/n! \neq 0 \quad \text{for any} \quad z \in D(0; R).$$

12. Prove that, for $n = 3, 4, \ldots$, the polynomial $z^n + nz - 1$ has n zeros inside the circle with centre at 0 and radius $1 + \sqrt{[2/(n-1)]}$.

Exercise 13 provides an alternative proof of Rouché's theorem, 7.7, making use of the concept of index introduced in 4.16.

13. Suppose f, g, and γ satisfy the conditions of Rouché's theorem, 7.7. Define F by $F(z) = [f(z) + g(z)]/f(z)$, and let Γ be the path $F \circ \gamma$. Prove that $\Gamma^* \subseteq D(1; 1)$ and hence show that $n(\Gamma, 0) = 0$. By applying Theorem 7.6 to F, deduce Rouché's theorem.

8 Applications of contour integration

This chapter is devoted to applications of Cauchy's residue theorem to the evaluation of definite integrals and the summation of series. The method will handle quite baroque examples; if we seem to have included some examples of this sort, it is because they provide valuable technical experience. Some contour integrals of particular importance are evaluated in Chapter 9. Very few of the integrals we consider can easily be treated by more elementary methods such as substitution, though numerical techniques may, of course, be available.

Improper and principal-value integrals

We begin with some technicalities, since it is necessary to clarify what our integrals mean. Except in our preview example, 7.1, we have so far only handled integrals of bounded functions on bounded intervals. Now that we come to consider integrals over infinite ranges and integrals of unbounded functions the differences between a Riemann approach and a Lebesgue approach become more marked. Riemann integrals are only defined directly for bounded functions on closed bounded intervals. Relaxation of these boundedness requirements can only be achieved by taking a limit of 'proper' Riemann integrals. The same problem does not arise with Lebesgue integrals, but we nonetheless need to consider a few functions which do not have 'proper' Lebesgue integrals. We therefore introduce, for either theory, improper integrals of various kinds, defined by limits. Further information on these integrals can be found in Binmore [4], or Weir [8].

8.1 Definitions

(1) Suppose the (real- or complex-valued) function f is integrable on every interval $[0, R]$ ($0 < R < \infty$). The *improper integral of* f

over $[0, \infty)$ is defined to be

$$\lim_{R \to \infty} \int_0^R f(x)\, dx,$$

if this limit exists.

(2) Suppose f is integrable on every closed bounded subinterval of $\mathbb{R}$. The *improper integral of f over* $\mathbb{R}$ is defined to be

$$\lim_{R,S \to \infty} \int_{-S}^R f(x)\, dx,$$

if this limit exists; here R and S tend to ∞ independently.

(3) Suppose f is integrable on every compact subinterval of $\mathbb{R}$. The *principal-value integral of f over* $\mathbb{R}$ is defined to be

$$\text{PV} \int_{-\infty}^{\infty} f(x)\, dx := \lim_{R \to \infty} \int_{-R}^R f(x)\, dx,$$

if this limit exists.

The reason for this proliferation of definitions is revealed in 8.2.

We write $\int_0^\infty f(x)\, dx$ and $\int_{-\infty}^\infty f(x)\, dx$ to mean, for a Riemann approach, the improper integral defined in, respectively, (1) and (2) above. Followers of Lebesgue should interpret $\int_0^\infty f(x)\, dx$ and $\int_{-\infty}^\infty f(x)\, dx$ as Lebesgue integrals where such exist, and as improper integrals otherwise.

8.2 Remarks

(1) Principal-value and improper integrals arise naturally from limits of contour integrals. Consider for example the integral of $f(z)$ round the contours $\tilde{\gamma}$ and $\tilde{\tilde{\gamma}}$ shown in Fig. 8.1: $\int_{\tilde{\gamma}} f(z)\, dz$ incorporates $\int_{-R}^R f(x)\, dx$ and $\int_{\tilde{\tilde{\gamma}}} f(z)\, dz$ incorporates $\int_{-S}^R f(x)\, dx$.

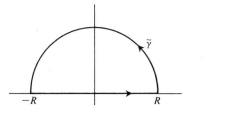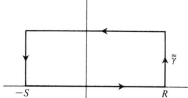

Fig. 8.1

(2) **Principal-value and improper integrals compared**   Trivially, if the improper integral of f over $\mathbb{R}$ exists, then so does the principal-value integral, and the two coincide. The converse fails: $\int_{-S}^{R} x \, dx = \frac{1}{2}(R^2 - S^2)$, which does not tend to a limit as R and S tend to ∞ independently, but $\int_{-R}^{R} x \, dx = 0$ for all R, so PV $\int_{-\infty}^{\infty} x \, dx$ exists and equals zero.

We conclude that the existence of the improper integral must be checked before PV $\int_{-\infty}^{\infty} f(x) \, dx$ may be replaced by $\int_{-\infty}^{\infty} f(x) \, dx$ (and it is customary to remove the PV-symbol wherever possible). A useful sufficient condition for f to have an improper integral over $\mathbb{R}$ [in fact a Lebesgue integral too] is

(i) f is integrable on compact subintervals of $\mathbb{R}$, and
(ii) $f(x) = O(|x|^{-p})$ for large $|x|$, for some constant $p > 1$.

[(3) **Improper and Lebesgue integrals compared**   The Dominated convergence theorem implies that if f has a Lebesgue integral on $[0, \infty)$ (or on $\mathbb{R}$), then f has an improper integral, and the two integrals coincide. The converse fails: the classic example is provided by $f(x) = x^{-1} \sin x$, which is not Lebesgue integrable on $[0, \infty)$ but does have an improper integral; see Weir [8], p. 100.]

The next four sections illustrate, by means of worked examples, methods for evaluating a variety of definite integrals. They are followed by a section which supplements the brief notes interspersed between examples and which attempts to give general guidelines on procedure. Further examples, involving Fourier and Laplace transforms, appear in Chapter 9. We note also that we already have a method for handling certain integrals of the form

$$\int_0^{2\pi} P(\cos \theta, \sin \theta) \, d\theta,$$

by converting them into integrals round the unit circle and applying the Residue theorem. Exercises 7.3 and 7.4 provide typical examples.

Integrals involving functions with a finite number of poles

We consider integrals of the form

$$\int_0^{\infty} \phi(x) \begin{Bmatrix} \sin mx \\ \cos mx \end{Bmatrix} dx \quad \text{and} \quad \int_{-\infty}^{\infty} \phi(x) \begin{Bmatrix} \sin mx \\ \cos mx \end{Bmatrix} dx,$$

where ϕ is a rational function, and PV-integrals of this type. The same methods apply when e^{imx} replaces the trigonometric factor; see 9.19(1).

Our first example is used to bring together the techniques developed for solving Example 7.1.

8.3 Example

To evaluate $\displaystyle\int_0^\infty \frac{1}{(x^2+1)^2(x^2+4)}\,dx.$

Solution. We integrate $f(z) = 1/[(z^2+1)^2(z^2+4)]$ round the contour $\gamma = \Gamma(0; R)$ shown in Fig. 8.2 (choosing $R > 2$). The function f is holomorphic inside and on γ except for a double pole at i and a simple pole at 2i. By Cauchy's residue theorem,

$$\int_{-R}^R f(x)\,dx + \int_{\Gamma_R} f(z)\,dz = 2\pi i[\mathrm{res}\{f(z); i\} + \mathrm{res}\{f(z); 2i\}].$$

We have, by 7.11(1),

$$\mathrm{res}\{f(z); i\} = \left[\frac{d}{dz}\frac{1}{(z+i)^2(z^2+4)}\right]_{z=i}$$

$$= \left[\frac{-2z(z+i) - 2(z^2+4)}{(z+i)^3(z^2+4)^2}\right]_{z=i} = \frac{-i}{36}$$

and, by 7.10(2),

$$\mathrm{res}\{f(z); 2i\} = \left[\frac{1}{(z^2+1)^2(2z)}\right]_{z=2i} = \frac{-i}{36}.$$

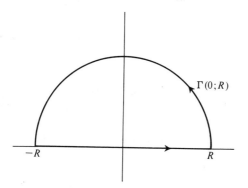

Fig. 8.2

Also

$$\left| \int_{\Gamma_R} f(z)\, dz \right| \le \int_0^\pi \frac{1}{(R^2-1)^2(R^2-4)}\, R\, d\theta = O(R^{-5}),$$

and

$$\int_{-R}^R f(x)\, dx = 2 \int_0^R \frac{1}{(x^2+1)^2(x^2+4)}\, dx.$$

Hence, letting $R \to \infty$, we get

$$\int_0^\infty \frac{1}{(x^2+1)^2(x^2+4)}\, dx = \frac{\pi}{18}. \qquad \square$$

Note We would not have been able to obtain the integral of $f(x)$ over $[0, \infty)$ using the contour $\Gamma(0; R)$ if f had not been such that $f(x) = f(-x)$ for all x. Notice how converting the range of integration from $[-R, R]$ to $[0, R]$ before taking the limit prevents an unnecessary principal value intruding.

8.4 Example

To evaluate $\displaystyle\int_{-\infty}^\infty \frac{\cos x}{x^2+x+1}\, dx.$

Solution. Integrate $f(z) = (z^2+z+1)^{-1} e^{iz}$ round $\gamma = \Gamma(0; R)$, where $\Gamma(0; R)$ is as in Fig. 8.2 and $R > 1$. The real part of $f(z)$, when z is real, is the required integrand. The function f is holomorphic inside and on γ except for a simple pole at $\omega = e^{2\pi i/3}$. By Cauchy's residue theorem and 7.10(2),

$$\int_{-R}^R f(x)\, dx + \int_{\Gamma_R} f(z)\, dz = 2\pi i\, \mathrm{res}\{f(z); \omega\}$$

$$= 2\pi i\, \frac{e^{i\omega}}{2\omega+1} = \frac{2\pi}{\sqrt{3}} e^{i(-\frac{1}{2}+\frac{1}{2}i\sqrt{3})}.$$

As in 7.16,

$$\left| \int_{\Gamma_R} f(z)\, dz \right| \le \int_0^\pi \frac{R e^{-R \sin \theta}}{|R^2 e^{2i\theta} + R e^{i\theta} + 1|}\, d\theta = O(R^{-1}).$$

Hence, letting $R \to \infty$ and equating real parts in the equation above, we obtain

$$PV \int_{-\infty}^\infty \frac{\cos x}{x^2+x+1}\, dx = \frac{2\pi}{\sqrt{3}} (\cos \tfrac{1}{2}) e^{-\frac{1}{2}\sqrt{3}}.$$

Since the integrand is $O(x^{-2})$ for large $|x|$, the improper integral [Lebesgue integral] exists and the PV-symbol may be deleted. □

Note When choosing f, one might have been tempted to take

$$f(z) = \frac{\cos z}{z^2 + z + 1}.$$

However this function does not behave suitably on the semicircular arc Γ_R: near $z = iR$, $\cos z$ behaves like e^R. A similar problem would arise if we chose to integrate $(z^2 + z + 1)^{-1} e^{iz}$ round a semicircle lying in the lower half-plane rather than the upper half-plane. This point comes up again in Example 9.19(1).

Each of the next two examples deals with an integral for which the semicircular contour $\Gamma(0; R)$ would seem appropriate, were it not that the most natural function to use has a pole on this contour (a simple one in 8.5, a multiple one in 8.6).

8.5 Example

To evaluate the improper integral $\displaystyle\int_0^\infty \frac{\sin x}{x}\, dx$.

Solution. Integrate $f(z) = e^{iz}/z$ round the contour γ shown in Fig. 8.3; f is holomorphic except for a simple pole at 0, which this indented semicircle avoids.

By Cauchy's theorem, and 3.5(1),

$$\int_{-R}^{-\varepsilon} f(x)\, dx - \int_{\Gamma_\varepsilon} f(z)\, dz + \int_{\varepsilon}^{R} f(x)\, dx + \int_{\Gamma_R} f(z)\, dz = 0.$$

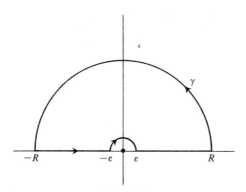

Fig. 8.3

We use 7.17(1) to calculate the contribution from Γ_ε in the limit as $\varepsilon \to 0$. It gives

$$\lim_{\varepsilon \to 0} \int_{\Gamma_\varepsilon} f(z)\, dz = i(\pi - 0)\mathrm{res}\{f(z); 0\} = i\pi.$$

By Jordan's inequality, 7.14(3), as in 7.16,

$$\left| \int_{\Gamma_R} f(z)\, dz \right| \leq 2 \int_0^{\frac{1}{2}\pi} e^{-R\sin\theta}\, d\theta \leq 2 \int_0^{\frac{1}{2}\pi} e^{-2R\theta/\pi}\, d\theta = O(R^{-1}).$$

Hence, letting $R \to \infty$ and $\varepsilon \to 0$,

$$i\pi = \lim_{R \to \infty, \varepsilon \to 0} \left[\int_{-R}^{-\varepsilon} \frac{e^{ix}}{x}\, dx + \int_\varepsilon^R \frac{e^{ix}}{x}\, dx \right] = \lim_{R \to \infty, \varepsilon \to 0} 2i \int_\varepsilon^R \frac{\sin x}{x}\, dx.$$

Therefore

$$\int_0^\infty \frac{\sin x}{x}\, dx = \tfrac{1}{2}\pi. \qquad \square$$

8.6 Example

To evaluate $\displaystyle\int_0^\infty \frac{x - \sin x}{x^3}\, dx$.

Solution. The integrand is the real part, when $z = x$ is real, of $k(z) = z^{-3}(z + ie^{iz})$. The Laurent expansion

$$z^{-3}(z + ie^{iz}) = z^{-3}\{z + i[1 + iz + \tfrac{1}{2}(iz)^2 + \ldots]\}$$

reveals a triple pole at 0. We cannot proceed exactly as in 8.5 because 7.17(1) is not applicable to k. However,

$$f(z) = z^{-3}(z + ie^{iz} - i)$$

is also such that its real part, when $z = x$ is real, is the required integrand, and f has only a simple pole at 0 (of residue $-\tfrac{1}{2}i$). Integrate f round the contour γ shown in Fig. 8.3. By Cauchy's theorem,

$$\int_{-R}^{-\varepsilon} x^{-3}(x + ie^{ix} - i)\, dx - \int_{\Gamma_\varepsilon} f(z)\, dz$$

$$+ \int_\varepsilon^R x^{-3}(x + ie^{ix} - i)\, dx + \int_{\Gamma_R} f(z)\, dz = 0.$$

The integrals over $[-R, -\varepsilon]$ and $[\varepsilon, R]$ combine to give

$$2 \int_\varepsilon^R \frac{x - \sin x}{x^3}\, dx;$$

doubling up now rather than equating real parts at the end avoids an unwanted principal value (cf. 8.3). Also

$$\left| \int_{\Gamma_R} f(z) \, dz \right| \leqslant \int_0^\pi \frac{R + e^{-R \sin \theta} + 1}{R^3} \, R \, d\theta = O(R^{-1}),$$

and $\lim_{\varepsilon \to 0} \int_{\Gamma_\varepsilon} f(z) \, dz = i(\pi - 0) \mathrm{res}\{f(z); 0\} = \frac{1}{2}\pi$ (by 7.17(1)). We now let $R \to \infty$ and $\varepsilon \to 0$ to get

$$\int_0^\infty \frac{x - \sin x}{x^3} \, dx = \frac{\pi}{4}. \qquad \Box$$

Integrals involving functions with infinitely many poles

The method of this section can be used to evaluate integrals of the type

$$\int_{-\infty}^\infty \phi(x) \begin{Bmatrix} \cos mx \\ \sin mx \end{Bmatrix} dx \quad \text{and} \quad \int_{-\infty}^\infty \phi(x) e^{imx} \, dx,$$

where $\phi(z)$ denotes a function, such as $\mathrm{cosec}\, z$, $\mathrm{sech}\, z$ or $(1 - e^z)^{-1}$, which has an infinite number of regularly spaced poles.

8.7 Example

To evaluate $\displaystyle\int_{-\infty}^\infty \frac{e^{ax}}{\cosh x} \, dx \ (-1 < a < 1).$

Solution. The function $f(z) = e^{az} \, \mathrm{sech}\, z$ has simple poles at $z = \frac{1}{2}(2n+1)\pi i \ (n \in \mathbb{Z})$. It is holomorphic inside and on the contour γ shown in Fig. 8.4 except for a simple pole inside γ, of residue

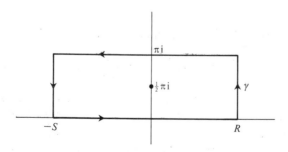

Fig. 8.4

$-ie^{\frac{1}{2}a\pi i}$ (by 7.10(2)). By Cauchy's residue theorem,

$$\int_{-S}^{R} \frac{e^{ax}}{\cosh x} \, dx + \int_{0}^{\pi} \frac{e^{a(R+iy)}}{\cosh(R+iy)} i \, dy + \int_{R}^{-S} \frac{e^{a\pi i}e^{ax}}{\cosh(x+\pi i)} \, dx$$

$$+ \int_{\pi}^{0} \frac{e^{a(-S+iy)}}{\cosh(-S+iy)} i \, dy = 2\pi e^{\frac{1}{2}a\pi i}.$$

Denote the second integral by I and the fourth by J. Then

$$|I| \leq \int_{0}^{\pi} \frac{2e^{aR}}{|e^{(R+iy)} + e^{-(R+iy)}|} \, dy \leq \int_{0}^{\pi} \frac{2e^{aR}}{|e^{R} - e^{-R}|} \, dy,$$

so $I \to 0$ as $R \to \infty$, since $a < 1$. Similarly,

$$|J| \leq \int_{0}^{\pi} \frac{2e^{-aS}}{|e^{-S} - e^{S}|} \, dy,$$

so $J \to 0$ as $S \to \infty$, since $a > -1$.

Letting R and S tend to ∞ independently, we obtain

$$\int_{-\infty}^{\infty} \frac{e^{ax}}{\cosh x} \, dx = \frac{2\pi e^{\frac{1}{2}a\pi i}}{1 + e^{a\pi i}} = \frac{2\pi}{e^{-\frac{1}{2}a\pi i} + e^{\frac{1}{2}a\pi i}} = \pi \sec \tfrac{1}{2}\pi a. \qquad \square$$

Notes The following observations motivate the choice of contour.

(i) Use of a semicircular contour would have involved an infinite sum of residues.

(ii) The vertical sides of the rectangle are along $x = R$ and $x = -S$, not $x = \pm R$, to avoid an unnecessary principal value.

(iii) The horizontal sides must avoid the poles. One is along $y = 0$, to yield the required integral; the other is along $y = \pi$, and yields a multiple of the required integral, because $\cosh(x + \pi i) = -\cosh x$ and $e^{a(x+\pi i)} = e^{ax}e^{a\pi i}$. To have been faced with an unrelated unknown integral would have been tiresome!

Deductions from known integrals

From certain known integrals it is possible to obtain the values of related integrals, as we illustrate in Example 8.9. A ubiquitous integral is $\int_{0}^{\infty} e^{-x^2} \, dx$, whose value is well known to be $\frac{1}{2}\sqrt{\pi}$. This is usually shown by methods other than contour integration, but with a bit more ingenuity than our previous examples have demanded, it can be derived from the Residue theorem. For completeness, and to provide variety, we outline the proof.

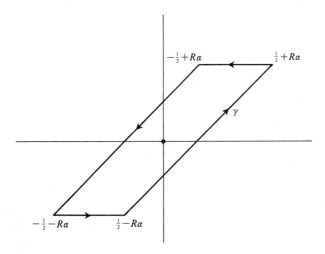

Fig. 8.5

8.8 Example

To prove $\int_0^\infty e^{-x^2}\, dx = \tfrac{1}{2}\sqrt{\pi}$.

Solution. Let $\alpha = e^{i\pi/4}$. We integrate $f(z) = (\operatorname{cosec} \pi z)e^{i\pi z^2}$ round the contour γ in Fig. 8.5, obtaining $\int_\gamma f(z)\, dz = 2\pi i\, \mathrm{res}\{f(z); 0\} = 2i$. On the slanting sides, we have $z = t\alpha \pm \tfrac{1}{2}$ $(-R \leqslant t \leqslant R)$ and $f(z) = \pm(\sec \pi t\alpha)e^{i\pi(it^2 \pm t\alpha + \frac{1}{4})}$, so that their combined contribution to $\int_\gamma f(z)\, dz$ is

$$\int_{-R}^{R} \alpha(e^{i\pi(it^2 + t\alpha + \frac{1}{4})} + e^{i\pi(it^2 - t\alpha + \frac{1}{4})})(\sec \pi t\alpha)\, dt$$

$$= 2i \int_{-R}^{R} e^{-\pi t^2}\, dt = \frac{4i}{\sqrt{\pi}} \int_0^{R\sqrt{\pi}} e^{-x^2}\, dx.$$

On the horizontal sides, $z = \pm R\alpha + t$ $(-\tfrac{1}{2} \leqslant t \leqslant \tfrac{1}{2})$, and an estimate of their contribution to the integral is given by

$$\left| \int_{-\frac{1}{2}}^{\frac{1}{2}} \frac{e^{i\pi(i\alpha^2 \pm 2R\alpha t + t^2)}}{\sin \pi(\pm R\alpha + t)}\, dt \right| \leqslant \int_{-\frac{1}{2}}^{\frac{1}{2}} \frac{2e^{-\pi R^2 \mp R\pi t\sqrt{2}}}{e^{\pi R/\sqrt{2}} - e^{-\pi R/\sqrt{2}}}\, dt.$$

Taking the limit as $R \to \infty$, the required result is obtained. ☐

8.9 Example

To evaluate the improper integral $\int_0^\infty \cos x^2\, dx$.

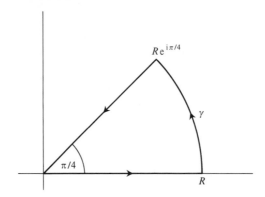

Fig. 8.6

Solution. Integrate $f(z) = e^{iz^2}$ round the contour γ shown in Fig. 8.6. By Cauchy's theorem,

$$\int_0^R e^{ix^2}\,dx + \int_0^{\frac{1}{4}\pi} e^{iR^2 e^{2i\theta}}Rie^{i\theta}\,d\theta + \int_R^0 e^{i(re^{\frac{1}{4}\pi i})^2}e^{\frac{1}{4}\pi i}\,dr = 0.$$

We have

$$\left|\int_0^{\frac{1}{4}\pi} e^{iR^2 e^{2i\theta}}Rie^{i\theta}\,d\theta\right| \leq R\int_0^{\frac{1}{4}\pi} e^{-R^2 \sin 2\theta}\,d\theta \leq R\int_0^{\frac{1}{4}\pi} e^{-4R^2\theta/\pi}\,d\theta,$$

and this is bounded by $\frac{1}{4}\pi R^{-1}(1 - e^{-R^2})$ (since, by Jordan's inequality, 7.14(3), $\sin 2\theta \geq 4\theta/\pi$ for $\theta \in [0, \frac{1}{4}\pi]$). Hence, letting $R \to \infty$,

$$\int_0^\infty e^{ix^2}\,dx = \frac{(1+i)}{\sqrt{2}}\int_0^\infty e^{-r^2}\,dr = \frac{(1+i)\sqrt{\pi}}{2\sqrt{2}}$$

(by 8.8). Equating real parts

$$\int_0^\infty \cos x^2\,dx = \sqrt{(\pi/8)}. \qquad \square$$

Note Ostensibly this example is solved by making a complex substitution in $\int_0^\infty e^{-x^2}\,dx$ of $xe^{i\pi/4}$ for x. This is legitimate if, as above, it is used in conjunction with Cauchy's theorem. Consider, for comparison, the following blatantly fallacious argument. Let $I = \int_0^\infty (1 + x^4)^{-1}\,dx$. Put $x = iy$. Then $I = \int_0^\infty (1 + y^4)^{-1}i\,dy$. So $I = 0$.

Integrals involving multifunctions

We may consider integrals of the form

$$\int_0^\infty \phi(x)\log x\,dx \quad \text{and} \quad \int_0^\infty \phi(x)x^{a-1}\,dx \quad (a > 0),$$

where $\phi(z)$ is meromorphic. Since logarithms and non-integer powers are multifunctions, we work in a cut plane with a specified branch of the multifunction being used. The branch point at 0 is avoided by means of an indentation. Recall the discussion on pp. 99–100 about integration in cut planes; this ensures that use of 4.6 or 7.4 is legitimate. We give one example here; for others see 9.12(3) and 9.19(3).

8.10 Example

To evaluate $\displaystyle\int_0^\infty \frac{\log x}{1+x^2}\,dx.$

Solution. We cut the plane along $(-\infty, 0]$, and take the holomorphic branch of the logarithm given by $\log z = \log |z| + i\theta$, where $z = |z|\,e^{i\theta}$ and $-\pi < \theta \le \pi$. Then $f(z) = (1+z^2)^{-1}\log z$ is holomorphic in the cut plane except for simple poles at $\pm i$. Let γ be the contour in Fig. 8.7. On the top edge of the cut, $\theta = \pi$, so $\log z = \log x + i\pi$, where $-z = x > 0$. Cauchy's residue theorem gives

$$\int_\varepsilon^R \frac{\log x}{1+x^2}\,dx + \int_{\Gamma_R} f(z)\,dz + \int_R^\varepsilon \frac{\log x + i\pi}{1+x^2}(-dx)$$

$$- \int_{\Gamma_\varepsilon} f(z)\,dz = 2\pi i\,\mathrm{res}\{f(z); i\}.$$

We have

$$\mathrm{res}\{f(z); i\} = \frac{\log i}{2i} = \frac{\frac{1}{2}\pi i}{2i}.$$

Also

$$\left| \int_{\Gamma_R} f(z)\,dz \right| \le \int_0^\pi \left| \frac{\log R + i\theta}{1 + R^2 e^{2i\theta}}\, Rie^{i\theta} \right| d\theta$$

$$\le \int_0^\pi \frac{(\log R + \pi)R}{R^2 - 1}\,d\theta = O(R^{-1}\log R)$$

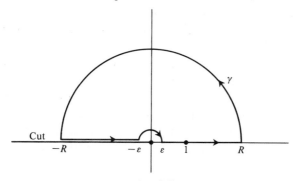

Fig. 8.7

and similarly

$$\left| \int_{\Gamma_\varepsilon} f(z)\, dz \right| \le \int_0^\pi \frac{(|\log \varepsilon| + \pi)\varepsilon}{1 - \varepsilon^2}\, d\theta = O(\varepsilon \log \varepsilon).$$

Invoke 7.15 to get, as $R \to \infty$ and $\varepsilon \to 0$,

$$2 \int_0^\infty \frac{\log x}{1 + x^2}\, dx + i\pi \int_0^\infty \frac{1}{1 + x^2}\, dx = \tfrac{1}{2}\pi^2 i.$$

By equating real parts we get

$$\int_0^\infty \frac{\log x}{1 + x^2}\, dx = 0. \qquad \square$$

(Equating imaginary parts gives us an integral we do not need contour integration to compute!)

Evaluation of definite integrals: summary

Many integrals can be handled in a variety of ways, but there are some principles which suggest the choice of function f and contour γ.

(1) $\int_\gamma f(z)\, dz$, or its real or imaginary part, should incorporate the required integral, or a quantity converging to it.

(2) Cauchy's theorem or Cauchy's residue theorem must be applicable, so f must be holomorphic, apart from poles. No poles may lie on γ. We may indent to avoid a simple pole; confronted with a multiple pole we try to redefine f. If a multifunction crops up, we work in a cut plane, with a suitably chosen holomorphic branch of the multifunction. The contour γ must not cross the cut, and indentations are made round branch points.

(3) If γ is the join of $\tilde{\gamma}$ and $\tilde{\tilde{\gamma}}$, and the required integral is to come from a limiting value of $\int_{\tilde{\gamma}} f(z)\, dz$, then $J := \int_{\tilde{\tilde{\gamma}}} f(z)\, dz$ must be manageable. We aim, for example, for:

(i) J tends to zero (as, for example, in 8.3) or to a finite non-zero limit (as in 8.5), or

(ii) J tends to a known integral (as in 8.9) or to a multiple of the required integral (as in 8.7).

(4) The sum of the residues of f inside γ must not be too awkward to compute (see 8.7 and Exercise 8.3).

(5) A semicircular contour is likely to produce a principal value integral. Ways of avoiding an otiose principal value are illustrated in 8.3, 8.6, and 8.7.

(6) Look before you leap! A preliminary substitution in a given real integral may greatly simplify the working, or even make contour integration unnecessary. Take, for example, the integral of $(1+x^3)^{-1}\sqrt{x}$ over $[0,\infty)$, which may be evaluated, painfully, using the 'keyhole' contour in Fig. 6.6. Alternatively, put $x^2 = y^3$ to obtain

$$\int_0^\infty \frac{\sqrt{x}}{1+x^3}\,dx = \frac{2}{3}\int_0^\infty \frac{1}{1+y^2}\,dy = \frac{2}{3}\Big[\tan^{-1}y\Big]_0^\infty = \frac{\pi}{3}.$$

Summation of series

Convergence tests establish that certain series converge to finite limits, but they do not yield the value of the sum. If an infinite sum can be recognized as a sum of residues of a meromorphic function, contour integration may enable us to evaluate it. Before discussing the general method we present a typical example.

8.11 Example

To prove that $\displaystyle\sum_{n=1}^\infty \frac{1}{n^2} = \frac{\pi^2}{6}$.

Solution. The function $f(z) = \pi z^{-2}\cot \pi z$ is holomorphic except for simple poles at n ($n = \pm1, \pm2, \ldots$) of residue $1/n^2$ (by 7.10(2)) and a triple pole at 0 of residue $-\pi^2/3$ (by 7.12(5)). Integrate f round the square contour γ_N shown in Fig. 8.8; γ_N^* is the square S_N with vertices at $(\pm1\pm i)(N+\tfrac{1}{2})$. Note that the vertical sides avoid

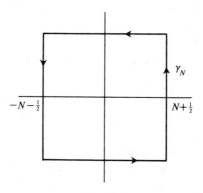

Fig. 8.8

the poles of f. By Cauchy's residue theorem,

$$\int_{\gamma_N} f(z)\,dz = 2\pi i\left(2\sum_{n=1}^{N}\frac{1}{n^2} - \frac{\pi^2}{3}\right).$$

It is now enough to show that $\int_{\gamma_N} f(z)\,dz \to 0$ as $N \to \infty$. We have

$$\left|\int_{\gamma_N} f(z)\,dz\right| \leq \sup_{z \in S_N}\left|\frac{\pi \cot \pi z}{z^2}\right| \times \text{length}(S_N)$$

$$\leq \sup_{z \in S_N}|\cot \pi z|\,\frac{4(2N+1)\pi}{(N+\frac{1}{2})^2},$$

which, by Lemma 8.12 below, is $O(N^{-1})$. □

8.12 Lemma

Let S_N be the square with vertices at $(\pm 1 \pm i)(N+\frac{1}{2})$ $(N = 1, 2, \ldots)$. Then there exists a constant C such that

$$|\cot \pi z| \leq C \quad \text{for all } N \text{ and for all } z \in S_N.$$

Proof. On the horizontal sides $z = x \pm i(N+\frac{1}{2})$,

$$|\cot \pi z| = \left|\frac{e^{i\pi[x \pm i(N+\frac{1}{2})]} + e^{-i\pi[x \pm i(N+\frac{1}{2})]}}{e^{i\pi[x \pm i(N+\frac{1}{2})]} - e^{-i\pi[x \pm i(N+\frac{1}{2})]}}\right|$$

$$\leq \frac{e^{\pi(N+\frac{1}{2})} + e^{-\pi(N+\frac{1}{2})}}{e^{\pi(N+\frac{1}{2})} - e^{-\pi(N+\frac{1}{2})}}$$

$$= \coth(N+\frac{1}{2}\pi)$$

$$\leq \coth \tfrac{3}{2}\pi \quad \text{(since } \coth t \text{ is decreasing for } t \geq 0\text{)}.$$

On the vertical sides $z = \pm(N+\frac{1}{2}) + iy$,

$$|\cot \pi z| = |\tan i\pi y| = |\tanh \pi y| \leq 1. \qquad \square$$

8.13 The summation of series by contour integration

The method used in Example 8.11 applies to any sum $\sum_{n=1}^{\infty} \phi(n)$, where ϕ is an even rational function holomorphic except at $\pm 1, \pm 2, \ldots$ and such that $\phi(z) = O(|z|^{-2})$ for large $|z|$: integrate $f(z) = \phi(z)\pi \cot \pi z$ round, as before, the contour γ_N in Fig. 8.8 and use Cauchy's residue theorem. Note that f has simple poles at n $(n = \pm 1, \pm 2, \ldots)$ of residue $\phi(n)$, and that the bound on ϕ and Lemma 8.12 ensure that $\int_{\gamma_N} f(z)\,dz \to 0$ as $N \to \infty$.

Under the same conditions on ϕ, $\sum_{n=1}^{\infty}(-1)^n\phi(n)$ can be evaluated by integrating $\phi(z)\pi \operatorname{cosec} \pi z$ round the same square contours. Here $\operatorname{res}\{f(z); n\} = (-1)^n \phi(n)$ $(n = \pm 1, \pm 2, \ldots)$. To prove the integrals tend to zero we need to know that $\operatorname{cosec} \pi z$ is bounded on the squares S_N by a constant independent of N. The proof of Lemma 8.12 can be easily modified to show that this is the case.

An adaptation of these techniques can be used to obtain series expansions of certain meromorphic functions: see Exercise 8.10 for an example.

Exercises

1. Prove (i) $\displaystyle\int_0^\infty \frac{1}{(x^2+a^2)(x^2+b^2)}\, dx = \frac{\pi}{2ab(a+b)}$ $\qquad (a, b > 0, \quad a \neq b)$,

   (ii) $\displaystyle\int_{-\infty}^\infty \frac{1}{(x^2+x+1)^2}\, dx = \frac{4\pi}{3\sqrt{3}}$, $\qquad$ (iii) $\displaystyle\int_{-\infty}^\infty \frac{1}{x^4+x^2+1}\, dx = \frac{\pi}{\sqrt{3}}$.

2. Prove (i) $\displaystyle\int_{-\infty}^\infty \frac{\cos x}{(x^2+a^2)}\, dx = \frac{\pi}{a} e^{-a}$ $\qquad (a > 0)$,

   (ii) $\displaystyle\int_0^\infty \frac{x^3 \sin x}{(x^2+1)^2}\, dx = \frac{\pi}{4e}$, $\qquad$ (iii) $\displaystyle\int_0^\infty \frac{\sin^2 x}{x^2}\, dx = \tfrac{1}{2}\pi$.

3. By integrating $(1+z^n)^{-1}$ round a suitable sector of angle $2\pi/n$, prove that, for $n = 2, 3, \ldots$,

$$\int_0^\infty (1+x^n)^{-1}\, dx = \frac{\pi}{n} \operatorname{cosec} \frac{\pi}{n}.$$

Evaluate also

$$\int_0^\infty x(1+x^n)^{-1}\, dx \quad (n = 3, 4, \ldots).$$

Could a semicircular contour have been used in either case?

4. Evaluate

$$\int_{-\infty}^\infty \frac{e^{ax}}{1+e^x}\, dx$$

and deduce that

$$\int_0^\infty \frac{x^{a-1}}{1+x}\, dx = \pi \operatorname{cosec} \pi a \quad (0 < a < 1).$$

Obtain the second integral directly by integrating a suitable function round the contour shown in Fig. 6.6.

5. Prove that (i) $\displaystyle\int_0^\infty \frac{\log x}{1+x^4}\, dx = -\frac{\pi^2}{8\sqrt{2}}$, $\qquad$ (ii) $\displaystyle\int_0^\infty \frac{(\log x)^2}{1+x^2}\, dx = \frac{\pi^3}{8}$.

6. 'The substitution $u = (a - ib)x$ gives, for $a > 0$ and $b \in \mathbb{R}$,

$$\int_0^\infty e^{-ax} e^{ibx}\, dx = \frac{1}{a - ib} \int_0^\infty e^{-u}\, du = \frac{a + ib}{a^2 + b^2}.$$

What is wrong with this argument? Give a correct derivation, by integrating e^{-z} round a suitable sector, and deduce the values of

$$\int_0^\infty e^{-ax} \cos bx \, dx, \qquad \int_0^\infty e^{-ax} \sin bx \, dx \quad (a > 0, \, b \in \mathbb{R}).$$

7. By integrating round a suitable sector, prove that

$$\int_0^\infty e^{-x^2} \sin(x^2) \, dx = \tfrac{1}{4}\sqrt{\pi}\sqrt{(\sqrt{2}-1)}.$$

8. Evaluate by contour integration

(i) $\displaystyle \int_{-\infty}^\infty \frac{\cos \pi x}{x^2 - 2x + 2} \, dx,$

(ii) $\displaystyle \int_0^\infty \frac{\cos ax - \cos bx}{x^2} \, dx \quad (a, \, b > 0),$

(iii) $\displaystyle \int_0^\infty \frac{x^2}{(1+x^2)^2} \, dx,$

(iv) $\displaystyle \int_0^\infty \frac{x^{\frac{1}{2}} \log x}{(1+x)^2} \, dx.$

9. In the plane cut along $[0, 1]$, take $[z(z-1)]^{\frac{1}{2}}$ to be the holomorphic branch of the square root which is real and positive at a given point $a > 1$. Prove that

$$2i \int_0^1 \frac{1}{[x(1-x)]^{\frac{1}{2}}(a-x)} \, dx = \int_\gamma \frac{1}{[z(z-1)]^{\frac{1}{2}}(a-z)} \, dz,$$

where $[x(1-x)]^{\frac{1}{2}}$ is positive on $[0, 1]$ and γ is a positively oriented contour enclosing 0 and 1 but not enclosing a. Hence show that

$$\int_0^1 \frac{1}{[x(1-x)]^{\frac{1}{2}}(a-x)} \, dx = \frac{\pi}{[a(a-1)]^{\frac{1}{2}}}.$$

10. Prove that

(i) $\displaystyle \sum_{n=-\infty}^\infty (n^2+1)^{-1} = \pi \coth \pi,$

(ii) $\displaystyle \sum_{n=1}^\infty (-1)^n (2n+1)^{-3} = \frac{\pi^3}{32}.$

11. By integrating

$$f(w) = \frac{\operatorname{cosec} w}{w(w-z)}$$

round a suitable contour, prove that

$$\operatorname{cosec} z = \frac{1}{z} - 2z \sum_{n=1}^\infty \frac{(-1)^n}{n^2\pi^2 - z^2} \quad (z \neq k\pi \, (k \in \mathbb{Z})).$$

12. Let the Taylor expansion of $\pi z \cot \pi z$ about 0 be $\sum_{n=0}^\infty a_n z^n$. Prove that

$$a_{2n} = -2 \sum_{k=1}^\infty k^{-2n} \quad (n \geq 1).$$

9 Fourier and Laplace transforms

On one level, integral transforms provide a versatile and systematic method for solving equations, on another they form the starting point for a rich theory having connections with many important branches of pure and applied mathematics. As explained in the preface, the emphasis in this chapter is on the part complex analysis plays in the rudiments of transform theory and its practical applications, rather than on the theory of transforms *per se*.

The motivating idea is a very simple one: if you cannot solve a given problem, transform it into one that you can, solve this, and then recapture the solution to the original problem. The examples in 9.21–9.25 (which have been grouped together for ease of reference) show such a process in action on ordinary and partial differential equations, and integral equations. The preceding sections set up the necessary machinery: the basic properties of Laplace and Fourier transforms are established, and methods for evaluating and inverting transforms are presented.

Technical apologia   Fourier and Laplace transforms are defined by infinite integrals. These may be taken to be Lebesgue integrals or (improper) Riemann integrals, and some readers will wish to opt for one interpretation, some for the other. We therefore introduce $\mathcal{I}(I)$, a class of functions which is differently defined for the Riemann and Lebesgue theories, and state results which are correct for either definition. This is feasible because, in deriving theorems about transforms, the calculations are essentially the same whichever integration is used. The differences come in the way the steps are justified once suitable conditions have been imposed on the functions. It is in the justification that Lebesgue integration scores heavily, with its powerful ammunition of the convergence theorems and the theorems of Fubini and Tonelli. For Riemann integrals the corresponding theorems are complicated to state and awkward to use; the zealous student will find the details in Apostol [1]. One escape route from this technical jungle should be mentioned at the outset. In deriving a solution to, for example, a differential equation, a step-by-step justification can often be avoided provided the putative solution is shown to work; see 9.20.

We shall use the following notation. Given $E \subseteq \mathbb{R}$, χ_E denotes the characteristic function defined by

$$\chi_E(x) = 1 \ (x \in E), \ \chi_E(x) = 0 \ (x \notin E).$$

9.1 Definition

Let I denote $\mathbb{R}$ or $[0, \infty)$.

For a Lebesgue approach, $\mathscr{I}(I)$ should be interpreted as $L^1(I)$, the (complex-valued) Lebesgue integrable functions on I.

For a Riemann approach, $\mathscr{I}(I)$ should be interpreted as the set of (complex-valued) piecewise continuous functions f such that each of f and $|f|$ has an improper integral on I; the terms used here were defined in 3.3 and 8.1. These conditions could be relaxed somewhat; see [1], Chapter 15.

The Laplace transform: basic properties and evaluation

We begin our discussion with the Laplace transform because at an elementary level it has a wider range of applications than the Fourier transform. For motivation, this section and the next may be read in parallel with the examples beginning at 9.21.

9.2 Definition

The *Laplace transform* of a function f defined on $[0, \infty)$ is given by

$$\bar{f}(p) = \int_0^\infty f(t) e^{-pt} \, dt.$$

It will be said to exist (for a given complex value of p) if and only if $f(t)e^{-pt} \in \mathscr{I}([0, \infty))$. Since $|f(t)e^{-pt}| = |f(t)|e^{-\mathrm{Re}\, pt}$ and any negative exponential function is integrable on $[0, \infty)$, the following conditions are sufficient (but not necessary) for the existence of $\bar{f}(p)$ for $\mathrm{Re}\, p > c$ (c a constant $\geqslant 0$):

(i) f is integrable on $[0, R]$ for $0 < R < \infty$, and
(ii) there exist constants M and T such that

$$|f(t)| \leqslant M e^{ct} \quad \text{for} \quad t \geqslant T.$$

We may think of the Laplace transform as an operator $\mathscr{L}$ taking a function f to its transform $\bar{f}$, so that $(\mathscr{L}f)(p) = \bar{f}(p)$. We follow convention in using p rather than z to denote the variable. We shall allow an abuse of notation and insert or omit the independent variables (t and p) as expedient. Specifically, it will be convenient to treat $\bar{f}(p)$ and $\mathcal{L}[f(t)]$ as synonymous.

Elementary calculations yield the following lemma.

9.3 Lemma

Provided the transforms involved exist
(1) for any constants a and b in $\mathbb{C}$, $\mathcal{L}[af(t)+bg(t)]= a\mathcal{L}[f(t)]+b\mathcal{L}[g(t)]$ (that is, $\mathcal{L}$ is linear);
(2) $\mathcal{L}[f(t/a)]=a\bar{f}(pa)$ $(a>0)$;
(3) $\mathcal{L}[e^{-at}f(t)]=\bar{f}(p+a)$ $(a\in\mathbb{C})$;
(4) $\mathcal{L}[f(t-a)H(t-a)]=e^{-ap}\bar{f}(p)$, where H is the Heaviside function defined by $H(t)=1$ $(t\geq 0)$, $H(t)=0$ $(t<0)$ (that is, $H=\chi_{[0,\infty)}$).

9.4 The Laplace transform of a derivative

Under appropriate conditions on f,

$$\mathcal{L}[f^{(n)}(t)]=p^n\bar{f}(p)-p^{n-1}f(0)-\ldots-f^{(n-1)}(0) \qquad (n=1,2,\ldots).$$

Sufficient conditions for the formula to be valid are:
  (i) $f, f', \ldots, f^{(n)}$ all exist, and their transforms exist,
  (ii) $f^{(n)}$ is continuous on $[0,\infty)$, and
  (iii) $f^{(k)}(t)e^{-pt}\to 0$ as $t\to\infty$ $(k=0,\ldots,n-1)$.
To derive the formula subject to (i)–(iii), we integrate by parts (which is certainly permissible if (i) and (ii) hold) to obtain

$$\mathcal{L}[f^{(n)}(t)] = \left[f^{(n-1)}(t)e^{-pt}\right]_0^\infty + p\int_0^\infty f^{(n-1)}(t)e^{-pt}\, dt$$

$$= -f^{(n-1)}(0)+p\mathcal{L}[f^{(n-1)}(t)] \quad \text{(by (iii))}.$$

The proof is completed by repeating this argument. □

Notice how the operator $\mathcal{L}$ converts derivatives of f into algebraic expressions involving $\bar{f}$. This is the crucial property involved in the solution of differential equations by transform methods.

9.5 The derivative of a Laplace transform

Suppose $\bar{f}(p)$ exists for Re $p>c$. Then $\bar{f}(p)$ is holomorphic for Re $p>c$, with derivatives given by differentiation under the integral sign:

$$\mathcal{L}[t^n f(t)]=(-1)^n\left(\frac{d}{dp}\right)^n\bar{f}(p).$$

Proof. Fix p such that $\operatorname{Re} p > c$ and write $\operatorname{Re} p - c = 2\eta$. Let h be such that $|h| < \eta$, so that $\operatorname{Re}(p+h) > c + \eta$. Then

$$\left| \frac{\bar{f}(p+h) - \bar{f}(p)}{h} + \int_0^\infty t f(t) e^{-pt}\, dt \right| = \left| \int_0^\infty f(t) e^{-pt} \left(\frac{e^{-ht} - 1}{h} + t \right) dt \right|$$

$$\leq \int_0^\infty \left| f(t) e^{-pt} \sum_{n=2}^\infty \frac{(th)^n}{n!\, h} \right| dt \qquad \text{(from the expansion for } e^{-ht})$$

$$\leq |h| \int_0^\infty \left| f(t) e^{-pt} \right| t^2 e^{t\,|h|}\, dt \leq |h| \int_0^\infty |f(t)| e^{-ct} t^2 e^{-\eta t}\, dt$$

and this tends to zero as $h \to 0$, since $t^2 e^{-\eta t}$ is bounded on $[0, \infty)$ and $|f(t)| e^{-ct} \in \mathcal{I}([0, \infty))$ by hypothesis. Hence $\bar{f}(p)$ is holomorphic for $\operatorname{Re} p > c$ and

$$\frac{d}{dp} \bar{f}(p) = -\mathcal{L}[t f(t)].$$

Higher-order derivatives are handled in the same way. $\qquad\square$

9.6 Elementary examples

Direct integration, combined with 9.3–9.5 enables a catalogue of basic transforms to be constructed. We record some of the most useful.

$f(t)$	$\bar{f}(p)$	Valid for
1	$1/p$	$\operatorname{Re} p > 0$
t^n	$n!/p^{n+1}$	$\operatorname{Re} p > 0,\ n = 1, 2, \ldots$
e^{-at}	$1/(p+a)$	$\operatorname{Re} p > -\operatorname{Re} a$
$\cos \omega t$	$p/(p^2 + \omega^2)$	$\operatorname{Re} p > |\operatorname{Im} \omega|$
$\sin \omega t$	$\omega/(p^2 + \omega^2)$	$\operatorname{Re} p > |\operatorname{Im} \omega|$

Table 9.1

The inversion of Laplace transforms

We next consider how a function f can be recovered from its transform $\bar{f}$. The simplest method is obviously 'inspection': recognition of a known transform. Unless one has recourse to a published table of transforms, this method is of limited use. More far-reaching (and the basis for the compilation of extensive tables)

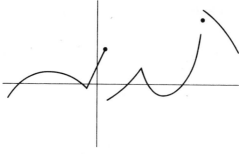

Fig. 9.1

is the Inversion theorem. The version we give is not the most general. The smoothness condition we impose is no hindrance in applications and facilitates the proof (though this is still sufficiently technical to be banished to an appendix to this chapter).

9.7 Piecewise smooth functions

We introduced piecewise continuous functions in 3.3. We note now that if f is piecewise continuous and is defined on a open interval containing t, then the left-hand and right-hand limits, $f(t-)$ and $f(t+)$, exist. Jump discontinuities of f occur at those points where $f(t-)$ and $f(t+)$ differ.

Given $I = \mathbb{R}$ or $[0, \infty)$, we say that a (real- or complex-valued) function f is *piecewise smooth on I* if f and f' are piecewise continuous on every closed bounded subinterval of I. This definition may appear somewhat ferocious, but piecewise smooth functions arise frequently and are easily recognized; see Fig. 9.1 for an archetypal example. Any continuously differentiable function is of course piecewise smooth.

9.8 The Inversion theorem for the Laplace transform

Suppose that f is piecewise smooth on $[0, \infty)$ and that $\bar{f}(p)$ exists for $\operatorname{Re} p > c \geqslant 0$. Then for $t > 0$,

$$\tfrac{1}{2}[f(t+) + f(t-)] = \frac{1}{2\pi i} \lim_{R \to \infty} \int_{\sigma - iR}^{\sigma + iR} \bar{f}(p) e^{pt} \, dp \qquad (\sigma > c).$$

The left-hand side simplifies to $f(t)$ if f is continuous at t.

The Inversion theorem guarantees that any continuous and piecewise smooth function is uniquely determined by its transform. It can in fact be proved that, for a function f which is merely continuous, $\bar{f} \equiv 0$ implies $f \equiv 0$ (Lerch's theorem). This uniqueness property is tacitly used whenever inverse transforms are obtained by inspection.

The inversion integral can frequently be evaluated by contour integration, using the techniques developed in Chapters 7 and 8. The following lemma gives a handy sufficient condition for an inverse Laplace transform to be a sum of residues.

9.9 Lemma

Let g be holomorphic except for a finite number of poles at $a_1, \ldots, a_n$, and suppose there exist positive constants M and k such that

$$|g(p)| \leq M\,|p|^{-k} \text{ for large } |p|.$$

Then, for $t > 0$ and $\sigma > \mathrm{Re}\ a_j\ (j = 1, \ldots, n)$,

$$\frac{1}{2\pi i} \lim_{R \to \infty} \int_{\sigma - iR}^{\sigma + iR} g(p) e^{pt}\, dp = \sum_{j=1}^{n} \mathrm{res}\{g(p) e^{pt};\, a_j\}.$$

Proof. We integrate $g(p)e^{pt}$ round the semicircular contour γ shown in Fig. 9.2 and apply Cauchy's residue theorem. On the semicircular arc ABC, $|p| \geq R - \sigma$ (by 1.4(3)), so the given bound on g implies, for large R,

$$\left| \int_{ABC} g(p) e^{pt}\, dp \right| \leq \int_{\pi/2}^{3\pi/2} M\,|R - \sigma|^{-k} e^{\sigma t - tR \cos \theta} R\, d\theta$$

$$= 2 \int_0^{\pi/2} M\,|R - \sigma|^{-k} e^{\sigma t - tR \sin \phi} R\, d\phi$$

$$\text{(putting } \phi = \theta - \tfrac{1}{2}\pi\text{),}$$

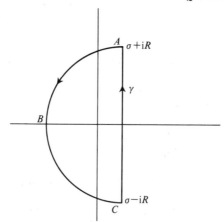

Fig. 9.2

and this tends to 0 as $R \to \infty$ (use Jordan's inequality if $0 < k \leqslant 1$; cf. 7.16). $\qquad \square$

Before turning to examples we give two further general results.

9.10 Theorem

Suppose that f satisfies the conditions for the Inversion theorem, 9.8, and that $\bar{f}(p)$ is expressible as

$$\bar{f}(p) = \sum_{n=0}^{\infty} a_n p^{-n-1},$$

where the series on the right hand side converges for $|p| > \rho$. Then, for $t > 0$,

$$\tfrac{1}{2}[f(t+) + f(t-)] = \sum_{n=0}^{\infty} a_n t^n / n!$$

(that is, term-by-term inversion is permissible).

Proof. Since $g(p) = \sum_{n=0}^{\infty} a_n p^{-n-1}$ converges for $|p| > \rho$, 2.8(3) implies that $\sum |a_n| r^{-n-1}$ converges for $r > \rho$. Thus
(1) $g(p)$ is holomorphic for $|p| > \rho$ (cf. 2.12),
(2) for $|p| \geqslant S > \rho$, $|g(p)| \leqslant |p|^{-1} \sum_{n=0}^{\infty} |a_n| S^{-n} = O(|p|^{-1})$, and
(3) on $|p| = S > \rho$, $|e^{pt}|$ is bounded, by K say, so that

$$|a_n p^{-n-1} e^{pt}| \leqslant K |a_n| S^{-n-1} = M_n,$$

and $\sum M_n$ converges.

By the Inversion theorem, with $\sigma > \rho$, we have

$$\tfrac{1}{2}[f(t+) + f(t-)] = \frac{1}{2\pi i} \lim_{R \to \infty} \int_{\sigma - iR}^{\sigma + iR} g(p) e^{pt} \, dp$$

$$= \frac{1}{2\pi i} \lim_{R \to \infty} \int_{\gamma} g(p) e^{pt} \, dp \quad \text{(taking } \gamma \text{ as shown in Fig. 9.3 and using (1), (2), and 9.8)}$$

$$= \frac{1}{2\pi i} \int_{\gamma(0;S)} g(p) e^{pt} \, dp \quad \text{(using (2) and 4.8; here } S \text{ is chosen such that } \rho < S < \sigma)$$

$$= \sum_{n=0}^{\infty} \frac{1}{2\pi i} a_n \int_{\gamma(0;S)} p^{-n-1} e^{pt} \, dp \quad \text{(using (3) and 3.13)}$$

$$= \sum_{n=0}^{\infty} a_n t^n / n! \quad \text{(by 5.4).} \qquad \square$$

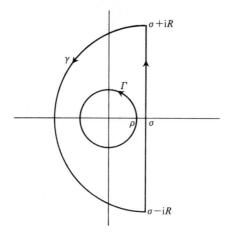

Fig. 9.3

The preceding results have contained a large component of complex analysis. The Convolution theorem, which enables us to invert a product of transforms, is really a piece of integration theory, and we accordingly relegate its proof to the appendix.

9.11 The Convolution theorem for the Laplace transform

Suppose f and g are such that $\bar{f}(p)$ and $\bar{g}(p)$ exist for $\operatorname{Re} p > c$. Then $\bar{f}(p)\bar{g}(p) = \bar{h}(p)$ for $\operatorname{Re} p > c$, where h is the *convolution* of f and g defined by

$$h(t) = \int_0^t f(x)g(t-x)\,dx \qquad (t \geqslant 0).$$

Summarizing the methods of inversion at our disposal, we have:
(1) inspection, backed up by 9.3–9.5;
(2) direct computation of the inversion integral in 9.8, usually aided by 9.9.
(3) term-by-term inversion, using 9.10;
(4) use of the Convolution theorem, 9.11, to invert a product of transforms.

9.12 Examples of inversion

(1) To find a function whose Laplace transform is $(p+1)/[p^2(p-1)]$.

Solution.

$$\frac{(p+1)}{p^2(p-1)} = \frac{2}{p-1} - \frac{(1+2p)}{p^2} = \mathscr{L}(2e^t - t - 2) \quad \text{(by 9.6),}$$

Alternatively, the Inversion theorem and Lemma 9.9 imply that the required inverse transform is the sum of the residues of

$$\frac{(p+1)}{p^2(p-1)} e^{pt}$$

at its poles at 1 (simple) and 0 (double). This sum is found, using 7.10(1) and 7.11(1), to be $2e^t - 2 - t$. ☐

(2) To find a function with Laplace transform $p/(p^2+\omega^2)^2$ (ω a constant).

Solution. Method 1:

$$\frac{p}{(p^2+\omega^2)^2} = \frac{1}{4i\omega}\left(\frac{1}{(p-i\omega)^2} - \frac{1}{(p+i\omega)^2}\right)$$

$$= \frac{1}{4i\omega}[\mathscr{L}(te^{i\omega t}) - \mathscr{L}(te^{-i\omega t})] \quad \text{(by 9.6 and 9.3(3) (or 9.5))}$$

$$= \mathscr{L}\left(\frac{t \sin \omega t}{2\omega}\right) \quad \text{(by 9.3(1)).}$$

Method 2:

$$\frac{p}{(p^2+\omega^2)^2} = -\frac{1}{2}\frac{d}{dp}\left(\frac{1}{p^2+\omega^2}\right) = \mathscr{L}\left(\frac{t \sin \omega t}{2\omega}\right) \quad \text{(by 9.5 and 9.6).}$$

Method 3: The Inversion theorem and Lemma 9.9 give

$$\mathscr{L}^{-1}\left(\frac{p}{(p^2+\omega^2)^2}\right) = \operatorname{res}\left\{\frac{pe^{pt}}{(p^2+\omega^2)^2} ; i\omega\right\} + \operatorname{res}\left\{\frac{pe^{pt}}{(p^2+\omega^2)^2} ; -i\omega\right\}$$

$$= \left[\frac{d}{dp}\left(\frac{pe^{pt}}{(p+i\omega)^2}\right)\right]_{p=i\omega} + \left[\frac{d}{dp}\left(\frac{pe^{pt}}{(p-i\omega)^2}\right)\right]_{p=-i\omega}$$

$$= \mathscr{L}\left(\frac{t \sin \omega t}{2\omega}\right) \quad \text{by a straightforward calculation.}$$

Method 4:

$$\frac{p}{(p^2+\omega^2)^2} = \sum_{n=0}^{\infty}(-1)^n(n+1)\frac{\omega^{2n}}{p^{2n+3}} \quad (|p|>|\omega|)$$

$$= \frac{1}{2\omega}\mathscr{L}\left(t\sum_{n=0}^{\infty}(-1)^n\frac{(\omega t)^{2n+1}}{(2n+1)!}\right) \quad \text{(by 9.10)}$$

$$= \left(\frac{t \sin \omega t}{2\omega}\right).$$

Method 5:

$$\frac{p}{(p^2+\omega^2)^2} = \frac{1}{\omega}\mathscr{L}(\cos \omega t)\mathscr{L}(\sin \omega t)$$

$$= \frac{1}{\omega}\mathscr{L}\left(\int_0^t \cos \omega x \sin \omega(t-x)\, dx\right) \quad \text{(by 9.11)}$$

$$= \mathscr{L}\left(\frac{t \sin \omega t}{2\omega}\right). \qquad \square$$

Note This example is designed to show off the variety of inversion techniques: for most functions the choice of methods is unlikely to be so great. Whatever method is used, it is usually easier to check the final answer by direct computation than to justify the use of 9.4, etc.

(3) To find a function whose Laplace transform is (a holomorphic branch of) $1/\sqrt{p}$.

Solution. We cut the plane along the negative real axis and take, for $p = |p|e^{i\theta} \ (-\pi < \theta \leqslant \pi)$, $\sqrt{p} = |p|^{\frac{1}{2}}e^{\frac{1}{2}i\theta}$. We wish to evaluate

$$\frac{1}{2\pi i}\lim_{R\to\infty}\int_{\sigma-iR}^{\sigma+iR}\frac{1}{\sqrt{p}}e^{pt}\, dp \qquad (\sigma > 0).$$

We use the 'keyhole' contour γ shown in Fig. 9.4. By Cauchy's theorem, the integral of $e^{pt}/\sqrt{p}$ round γ is zero. That the integrals along AB and EF tend to zero is proved exactly as in 9.9. Also

$$\left|\int_{\gamma(0;\,\varepsilon)}\frac{1}{\sqrt{p}}e^{pt}\, dp\right| \leqslant \int_0^{2\pi}\frac{1}{\sqrt{\varepsilon}}e^{\varepsilon t \cos\theta}\varepsilon\, d\theta = O(\varepsilon^{\frac{1}{2}}).$$

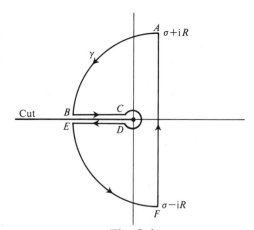

Fig. 9.4

On BC, $\sqrt{p} = i\sqrt{x}$ $(x > 0)$ and on DE, $\sqrt{p} = -i\sqrt{x}$ $(x > 0)$. The integrals along BC and DE combine to give

$$-2\int_\varepsilon^R \frac{1}{i\sqrt{x}} e^{-xt}\, dx.$$

Hence, letting $\varepsilon \to 0$ and $R \to \infty$,

$$\mathcal{L}^{-1}(1/\sqrt{p}) = \frac{1}{\pi}\int_0^\infty \frac{1}{\sqrt{x}} e^{-xt}\, dx$$

$$= \frac{2}{\pi\sqrt{t}}\int_0^\infty e^{-y^2}\, dy \quad \text{(putting } xt = y^2)$$

$$= 1/\sqrt{(\pi t)} \qquad \text{(by 8.8).} \qquad \square$$

The Fourier transform

9.13 Definition

The *Fourier transform* of $f \in \mathcal{S}(\mathbb{R})$ is defined, for all real s, by

$$(\mathcal{F}f)(s) = \hat{f}(s) = \int_{-\infty}^\infty f(x)e^{-isx}\, dx.$$

Variants on this appear in some books: e^{isx} instead of e^{-isx} or an inserted normalization factor of $\sqrt{(1/2\pi)}$.

The Laplace transform can be regarded as a special case of the Fourier transform, a fact we shall exploit in the appendix. Write $p = u + is$ and suppose $\bar{f}(p)$ exists. Let $g(x) = e^{-ux}f(x)\chi_{[0,\infty)}(x)$. Then

$$\hat{g}(s) = \int_0^\infty e^{-ux}f(x)e^{-isx}\, dx = \bar{f}(p).$$

The next three results parallel those in 9.3–9.5. We leave them as exercises, of varying technical difficulty. The proofs of the Inversion theorem, 9.17, and the Convolution theorem, 9.18, appear in the appendix.

9.14 Lemma

Provided all the transforms exist
(1) $\mathcal{F}[af(x) + bg(x)] = a\mathcal{F}[f(x)] + b\mathcal{F}[g(x)]$ $(a, b \in \mathbb{C})$;
(2) $\mathcal{F}[f(x/a)] = a\hat{f}(sa)$ $(a > 0)$;
(3) $\mathcal{F}[e^{-ixa}f(x)] = \hat{f}(s + a)$ $(a \in \mathbb{R})$.

9.15 The Fourier transform of a derivative

Under appropriate conditions on f,

$$\mathcal{F}[f^{(n)}(x)] = (\mathrm{i}s)^n \hat{f}(s).$$

This is derived by repeated integration by parts. Sufficient assumptions are: $f, f', \ldots, f^{(n)} \in \mathscr{S}(\mathbb{R})$ and $f^{(n)}$ is continuous. These are enough to legitimize integration by parts, and also to force, for each k, $f^{(k)}(x)\mathrm{e}^{-\mathrm{i}sx} \to 0$ as $|x| \to \infty$, and so dispose of the integrated terms.

9.16 The derivative of a Fourier transform

Under conditions on f sufficient to justify repeated differentiation under the integral sign,

$$\mathcal{F}[x^n f(x)] = \mathrm{i}^n \hat{f}^{(n)}(s).$$

9.17 The Inversion theorem for the Fourier transform

Let f be a piecewise smooth function in $\mathscr{S}(\mathbb{R})$. If

$$\hat{f}(s) = \int_{-\infty}^{\infty} f(x)\mathrm{e}^{-\mathrm{i}sx}\,\mathrm{d}x,$$

then

$$\tfrac{1}{2}[f(x+)+f(x-)] = \frac{1}{2\pi}\,\mathrm{PV}\int_{-\infty}^{\infty} \hat{f}(s)\mathrm{e}^{\mathrm{i}sx}\,\mathrm{d}s.$$

Note It is in general necessary to work with the PV-integral. Take $f(x) = \mathrm{e}^{-x}\chi_{[0,\infty)}(x)$. Then $\hat{f}(s) = (1+s^2)^{-1} - \mathrm{i}s(1+s^2)^{-1}$ and $\hat{f} \notin \mathscr{S}(\mathbb{R})$.

9.18 The Convolution theorem for the Fourier transform

Let f and g belong to $\mathscr{S}(\mathbb{R})$. Then for $s \in \mathbb{R}$, $\hat{f}(s)\hat{g}(s) = \hat{h}(s)$, where the *convolution* h is defined by

$$h(x) = \int_{-\infty}^{\infty} f(y)g(x-y)\,\mathrm{d}y.$$

9.19 Examples of Fourier transforms

Our examples, which illustrate the foregoing results and the techniques of contour integration, are drawn from probability

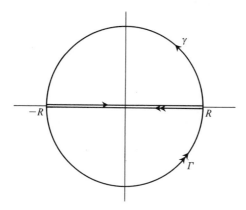

Fig. 9.5

theory: we compute the characteristic functions of some important probability distributions.

(1) (The Cauchy distribution) Let $f(x) = 1/(1+x^2)\pi$. Then $\hat{f}(s) = e^{-|s|}$.

Solution. The natural way to evaluate

$$\hat{f}(s) = \int_{-\infty}^{\infty} \frac{e^{-isx}}{\pi(1+x^2)}\,dx$$

is to integrate $g(z) := (1+z^2)^{-1}e^{-isz}/\pi$ round a semicircular contour. When $z = Re^{i\theta}$, $|e^{-isz}| = e^{sR\sin\theta}$, and this stays bounded as $R \to \infty$ if and only if $s \sin \theta \leq 0$. Accordingly we consider the cases $s \leq 0$ and $s > 0$ separately, and choose our contour to be γ in the first case and Γ in the second, where γ and Γ are as illustrated in Fig. 9.5. In either case the integral of $g(z)$ round the semicircular arc tends to 0 as $R \to \infty$ (cf. 7.16). Applying Cauchy's residue theorem and letting $R \to \infty$, we obtain (from 7.10)

$$\hat{f}(s) = \begin{cases} -2\pi i\,\text{res}\{g(z); i\} &= e^s \quad (s \leq 0), \\ -2\pi i\,\text{res}\{g(z); -i\} = e^{-s} \quad (s > 0), \end{cases}$$

whence $\hat{f}(s) = e^{-|s|}$ $(s \in \mathbb{R})$. $\qquad\qquad\square$

The inversion integral

$$\frac{1}{2\pi} \text{PV} \int_{-\infty}^{\infty} e^{-|s|} e^{isx}\,ds$$

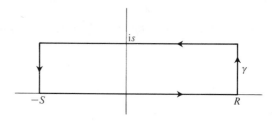

Fig. 9.6

may be directly computed. This is done by splitting the integrand into its real and imaginary parts and the range of integration into $(-\infty, 0]$ and $[0, \infty)$. We obtain, as the Inversion theorem leads us to expect, $1/(1+x^2)\pi$.

(2) (The normal distribution) Let

$$f(x) = \frac{1}{\sqrt{(2\pi)}} e^{-\frac{1}{2}x^2}.$$

Then $\hat{f}(s) = e^{-\frac{1}{2}s^2}$.

Solution.

$$\sqrt{(2\pi)}\hat{f}(s) = \int_{-\infty}^{\infty} e^{-\frac{1}{2}x^2 - isx}\, dx = e^{-\frac{1}{2}s^2} \int_{-\infty}^{\infty} e^{-\frac{1}{2}(x+is)^2}\, dx.$$

Integrate $e^{-\frac{1}{2}z^2}$ round the rectangle with vertices at $-S$, R, $R+is$, and $-S+is$ (shown in Fig. 9.6 in the case $s>0$). By Cauchy's theorem,

$$\int_{-S}^{R} e^{-\frac{1}{2}x^2}\, dx + \int_{0}^{s} e^{-\frac{1}{2}(R+iy)^2}i\, dy + \int_{R}^{-S} e^{-\frac{1}{2}(x+is)^2}\, dx + \int_{s}^{0} e^{-\frac{1}{2}(-S+iy)^2}i\, dy = 0.$$

Here

$$\left| \int_{0}^{s} e^{-\frac{1}{2}(R+iy)^2}i\, dy \right| \le e^{-\frac{1}{2}R^2} \int_{0}^{|s|} e^{\frac{1}{2}y^2}\, dy \to 0 \text{ as } R \to \infty,$$

and similarly

$$\int_{0}^{s} e^{-\frac{1}{2}(-S+iy)^2}i\, dy \to 0 \text{ as } S \to \infty.$$

Hence, letting R and S tend to ∞,

$$\int_{-\infty}^{\infty} e^{-\frac{1}{2}x^2}\, dx = \int_{-\infty}^{\infty} e^{-\frac{1}{2}(x+is)^2}\, dx.$$

The left hand side is $\sqrt{(2\pi)}$, by 8.8, so $\hat{f}(s) = e^{-\frac{1}{2}s^2}$. ☐

Note how contour integration makes the complex substitution of $x + is$ for x respectable (see the note following Example 8.9).

Symmetry shows the result obtained to be consistent with the Inversion theorem.

(3) (The gamma distribution) For $\lambda > 0$ and $t > 0$, let

$$f(x) = \frac{1}{\Gamma(t)} \lambda^t x^{t-1} e^{-\lambda x} \chi_{[0,\infty)}(x),$$

where

$$\Gamma(t) := \int_0^\infty x^{t-1} e^{-x} \, dx$$

defines the *gamma function*. Then

$$\hat{f}(s) = \left(\frac{\lambda}{\lambda + is} \right)^t,$$

where the right hand side is suitably defined.

Solution. When t is an integer we may integrate round a sector (motivated by the formal substitution of $(\lambda + is)x$ for x). In general, we have to contend with a multifunction. We work in the plane cut along the negative real axis and take $z^{t-1} = |z|^{t-1} e^{i\theta(t-1)}$ for $z = |z| e^{i\theta}$ $(-\pi < \theta \leqslant \pi)$. Integrate $g(z) = z^{t-1} e^{-z}$ round the contour indicated in Fig. 9.7. On CD, $z = (\lambda + is)u$ with $u > 0$, so

$$\int_{DC} g(z) \, dz = -(\lambda + is)^t \int_\varepsilon^R u^{t-1} e^{-(\lambda + is)u} \, du,$$

while

$$\int_{AB} g(z) \, dz = \int_\varepsilon^R x^{t-1} e^{-x} \, dx.$$

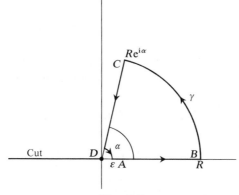

Fig. 9.7

Also

$$\left| \int_{BC} g(z)\, dz \right| \leqslant \int_0^{|\alpha|} |(Re^{i\theta})^{t-1} e^{-Re^{i\theta}} Rie^{i\theta}|\, d\theta \quad \text{(where } \tan\alpha = s/\lambda)$$

$$\leqslant \int_0^{|\alpha|} R^t e^{-R\cos\theta}\, d\theta$$

$$\leqslant |\alpha|\, R^t e^{-R\cos\alpha},$$

which tends to 0 as $R \to \infty$, and

$$\left| \int_{DA} g(z)\, dz \right| \leqslant |\alpha|\, \varepsilon^t e^{-\varepsilon\cos\alpha},$$

which tends to zero as $\varepsilon \to 0$, since $t > 0$. Apply Cauchy's theorem and take the limit as $R \to \infty$ and $\varepsilon \to 0$ to obtain

$$(\lambda + is)^t \int_0^\infty u^{t-1} e^{-(\lambda+is)u}\, du = \int_0^\infty x^{t-1} e^{-x}\, dx.$$

Thus

$$\hat{f}(s) = \left(\frac{\lambda}{\lambda+is}\right)^t$$

(equal to

$$\left|\frac{\lambda}{\lambda+is}\right|^t e^{it\alpha},$$

where $\tan\alpha = s/\lambda$ and $-\tfrac{1}{2}\pi < \alpha < \tfrac{1}{2}\pi$). $\qquad\qquad\square$

Applications to differential equations, etc.

In this section we illustrate the use of transforms in the solution of simple ordinary and partial differential equations and integral equations.

We show the idea behind the method with a very simple introductory example. Consider the equation

$$L\frac{dI}{dt} + RI = E \quad (t \geqslant 0), \quad \text{with } I = 0 \text{ when } t = 0,$$

in which $E, L,$ and R are constants. We multiply the given equation by e^{-pt} and integrate over $[0, \infty)$ to get (by linearity of $\mathscr{L}$),

$$L\int_0^\infty \frac{dI}{dt} e^{-pt}\, dt + R\int_0^\infty I(t)e^{-pt}\, dt = E\int_0^\infty e^{-pt}\, dt.$$

By 9.4 the first integral is $p\bar{I}(p) - I(0)$. Hence, using the given initial condition we have

$$Lp\bar{I}(p) + R\bar{I}(p) = \frac{E}{p}$$

whence

$$\bar{I}(p) = \frac{E}{p(R + Lp)} = \frac{E}{R}\left(\frac{1}{p} - \frac{1}{p + R/L}\right).$$

We recognize the right-hand side as the transform of $\frac{E}{R}(1 - e^{-Rt/L})$. This function is the unique solution for $I(t)$.

We can apply the same idea much more widely. Suppose t takes values in $[0, \infty)$ (in $\mathbb{R}$). Then the Laplace transform (the Fourier transform) will convert

(1) an ordinary differential equation for $f(t)$ with constant or polynomial coefficients into an algebraic equation or ordinary differential equation for $\bar{f}(p)$ ($\hat{f}(s)$) (Examples 9.21, 9.22, and Exercise 9.7);

(2) simultaneous ordinary differential equations with constant coefficients, in $f_1(t), \ldots, f_n(t)$, into simultaneous linear equations in $\bar{f}_1(p), \ldots, \bar{f}_n(p)$ ($\hat{f}_1(s), \ldots, \hat{f}_n(s)$) (Exercise 9.8);

(3) a partial differential equation for $u(x, t)$ (of suitable type) into an ordinary differential equation for $\bar{u}(x, p)$ ($\hat{u}(x, s)$), with x as the variable (Examples 9.23, 9.24 and Exercise 9.12).

These statements should not be treated as rules which are universally applicable. If the method is to be carried through successfully to solve the given equation(s), certain technical provisos must be added, and the right boundary and/or initial conditions must be given. The transform used must fit the problem. For example, the Laplace transform is only applicable where one variable has domain $[0, \infty)$, as, conveniently, is usually the case when time is involved. Sometimes the form of the boundary condition will restrict the choice of method, but often several methods are feasible: variants on the standard transforms (sine and cosine transforms, for instance) are more efficient for certain problems. We refer the reader who wishes to know exactly which classes of equations can be solved by which transforms to specialized textbooks. In (1) and (2), transforms are particularly suitable where the equations involve unknown constants or functions (often standing for physical quantities, for example, resistance, inductance, and capacitance in electric circuit problems). Equations involving only numerical constants and given functions are usually most easily solved by more elementary means; using a transform is applying a sledgehammer to a nut.

9.20 Important note

Before we embark on examples we amplify a comment made earlier. It is often relatively easy to *obtain* a transform solution, but if one does not establish the validity of each step in the calculation, one has no guarantee that that solution is correct. The following remarks may help the reader to avoid an excessively pedantic or excessively cavalier attitude.

(1) **Don't break the rules** For example, integrals must have integrable integrands, and the rules of contour integration must be obeyed (remember that integrals round arcs must tend to the limits claimed, and a cut plane should be used with a multifunction).

(2) **Amenable functions** Serious problems are likely to arise primarily with 'pathological' functions; solutions of differential equations are, a priori, reasonably smooth, and hence behave well under transforms. However one may have to impose restrictions on a function's rate of growth and limiting behaviour which are not built into the original problem. In many applied problems such technical restrictions are acceptable on physical grounds.

(3) **Don't make work** In many cases delicate analysis is neither necessary nor appropriate. It is frequently easier to proceed formally, and to check at the end that the solution does satisfy the equation and any additional conditions. Some familiarity with existence and uniqueness theorems is valuable here; one then knows what to expect by way of solutions.

9.21 Example (A linear ordinary differential equation)

To solve

$$f''(t) + f(t) = \begin{cases} \cos t & (0 \leqslant t \leqslant \pi), \\ 0 & (t > \pi), \end{cases}$$

given $f(0) = f'(0) = 0$.

Solution. We operate on the differential equation by $\mathscr{L}$ to get

$$\mathscr{L}[f''(t)] + \mathscr{L}[f(t)] = \int_0^\pi (\cos t) e^{-pt} \, dt = \frac{p(1 - e^{-p\pi})}{(1 + p^2)}.$$

Hence, using 9.4 and the initial conditions,

$$(1 + p^2)\bar{f}(p) = \frac{p(1 + e^{-p\pi})}{(1 + p^2)}.$$

Example 9.12(2) implies that $p/(1 + p^2)^2 = \mathscr{L}(\tfrac{1}{2}t \sin t)$. By 9.3(4),

$$f(t) = \tfrac{1}{2}t \sin t + \tfrac{1}{2}(t - \pi)\sin(t - \pi)\, H(t - \pi)$$

$$= \begin{cases} \tfrac{1}{2}t \sin t & (0 \leqslant t \leqslant \pi), \\ \tfrac{1}{2}\pi \sin t & (t > \pi). \end{cases} \qquad \square$$

Note We may consider, more generally,
$$f''(t) + f(t) = k(t) \ (t \geqslant 0) \text{ subject to } f(0) = f'(0) = 0.$$
The transformed equation is then
$$\bar{f}(p) = \bar{k}(p)/(1 + p^2),$$
whence, for any reasonably well behaved function k,
$$f(t) = \int_0^t k(x)\sin(t - x)\,\mathrm{d}x \qquad (t \geqslant 0),$$
by the Convolution theorem, 9.11.

9.22 Example (An ordinary differential equation with polynomial coefficients)

To find a solution of Bessel's equation of order zero, viz.
$$tf''(t) + f'(t) + tf(t) = 0.$$

Solution. Apply $\mathscr{L}$ and use 9.4 and 9.5. The equation transforms into
$$-\frac{\mathrm{d}}{\mathrm{d}p}[p^2\bar{f}(p) - pf(0) - f'(0)] + [p\bar{f}(p) - f(0)] - \frac{\mathrm{d}}{\mathrm{d}p}\bar{f}(p) = 0.$$
Hence
$$-(p^2 + 1)\frac{\mathrm{d}}{\mathrm{d}p}\bar{f}(p) = p\bar{f}(p),$$
which is satisfied by
$$\bar{f}(p) = A(p^2 + 1)^{-\frac{1}{2}} \qquad (A \text{ constant}),$$
where the right-hand side is defined so as to be holomorphic in the plane cut between $-i$ and i. For $|p| > 1$,
$$(p^2 + 1)^{-\frac{1}{2}} = \sum_{n=0}^{\infty} (-1)^n (2n)!/[(n!)^2 2^{2n} p^{2n+1}].$$
Theorem 9.10 now implies
$$f(t) = A \sum_{n=0}^{\infty} (-1)^n (\tfrac{1}{2}t)^{2n}/(n!)^2 =: AJ_0(t).$$
It can be checked, with the aid of 2.12, that this series does satisfy Bessel's equation. $\square$

Note Bessel's equation is of second order, so does have, as one would expect, two linearly independent solutions, $J_0(t)$ and $Y_0(t)$. However, only $J_0(t)$ is sufficiently well-behaved to be obtained via the Laplace transform; $Y_0(t)$ 'blows up' at $t = 0$.

9.23   Example (The diffusion equation)

To find the function $u(x, t)$ which is defined and continuous in $\{(x, t): x \geqslant 0, t \geqslant 0\}$ and which satisfies

(i) $\dfrac{\partial u}{\partial t} = \kappa \dfrac{\partial^2 u}{\partial x^2}$ (κ a constant > 0);

(ii) $u(x, 0) = 0$ $(x \geqslant 0)$;

(iii) $u(0, t) = U$ $(t \geqslant 0)$ (U a constant);

(iv) $u(x, t)$ remains bounded as $x \to \infty$.

Solution. We operate by $\mathscr{L}$ on the variable t, writing

$$\bar{u}(x, p) = \int_0^\infty u(x, t) e^{-pt} \, dt.$$

By 9.4 and (ii),

$$\int_0^\infty \frac{\partial}{\partial t} u(x, t) e^{-pt} \, dt = p\bar{u}(x, p).$$

Treating p as fixed and assuming differentiation under the integral sign with respect to x is permissible twice, (i) transforms into

$$p\bar{u}(x, p) = \kappa \frac{d^2 \bar{u}}{dx^2}.$$

This has solution

$$\bar{u}(x, p) = A(p) e^{x\sqrt{(p/\kappa)}} + B(p) e^{-x\sqrt{(p/\kappa)}},$$

where $A(p)$ and $B(p)$ are functions of p. Here we assume the plane cut along $(-\infty, 0]$ and take, for $p = re^{i\theta}$, $\sqrt{p}$ to be $r^{\frac{1}{2}} e^{\frac{1}{2}i\theta}$, where $-\pi < \theta \leqslant \pi$. Thus $\mathrm{Re}\,\sqrt{p} \geqslant 0$.

Now operate by $\mathscr{L}$ on (iii) and (iv). From (iii),

$$\bar{u}(0, p) = U/p \qquad (\mathrm{Re}\, p > 0),$$

while (iv) implies $\bar{u}(x, p)$ remains bounded as $x \to \infty$. Hence $A(p) = 0$ and $B(p) = U/p$, so

$$\bar{u}(x, p) = \frac{U}{p} e^{-x\sqrt{(p/\kappa)}} \qquad (\mathrm{Re}\, p > 0).$$

By the Inversion theorem, for $t > 0$,

$$u(x, t) = \frac{1}{2\pi i} \lim_{R \to \infty} \int_{\sigma - iR}^{\sigma + iR} \frac{U}{p} e^{pt - x\sqrt{(p/\kappa)}} \, dp \qquad (\sigma > 0).$$

To evaluate this we use the same 'keyhole' contour as in Example 9.12(3); see Fig. 9.4. For $p = re^{i\theta}$ $(-\pi < \theta \leqslant \pi)$,

$$\left| \frac{1}{p} e^{-x\sqrt{(p/\kappa)}} \right| = \frac{1}{r} e^{-x\sqrt{(r/\kappa)}\cos\frac{1}{2}\theta} = O(r^{-1}).$$

Hence, by Jordan's inequality (as in the proof of Lemma 9.9), the integrals along AB and EF tend to zero as $R \to \infty$. On the small circle of radius ε,

$$\frac{1}{p} e^{pt - x\sqrt{(p/\kappa)}} = \frac{1}{p} + h(p),$$

where $h(p) = O(1/\sqrt{\varepsilon})$. Hence

$$\int_{\gamma(0;\varepsilon)} \frac{1}{p} e^{pt - x\sqrt{(p/\kappa)}} \, dp = 2\pi i + O(\sqrt{\varepsilon}) \quad \text{(by 3.7(1) and 3.10)}.$$

On BC, $\sqrt{p} = i\sqrt{r}$ and on DE, $\sqrt{p} = -i\sqrt{r}$ $(r > 0)$. The function $(U/p) e^{pt - x\sqrt{(p/\kappa)}}$ is holomorphic inside and on the 'keyhole' so, applying Cauchy's theorem and letting $R \to \infty$ and $\varepsilon \to 0$, we get

$$u(x, t) = U - \frac{U}{2\pi i} \int_0^\infty e^{-rt + ix\sqrt{(r/\kappa)}} \frac{1}{r} \, dr + \frac{U}{2\pi i} \int_0^\infty e^{-rt - ix\sqrt{(r/\kappa)}} \frac{1}{r} \, dr$$

$$= U - \frac{U}{\pi} \int_0^\infty \sin[x\sqrt{(r/\kappa)}] e^{-rt} \frac{1}{r} \, dr$$

$$= U - \frac{2U}{\pi} \int_0^\infty \sin[xy/\sqrt{(2\kappa t)}] e^{-\frac{1}{2}y^2} \frac{1}{y} \, dy \quad \text{(putting } y^2 = 2rt\text{)}.$$

This integral can be expressed in terms of the error function

$$\mathrm{erf}(x) := \frac{2}{\sqrt{\pi}} \int_0^x e^{-y^2} \, dy,$$

as $u(x, t) = U\left[1 - \mathrm{erf}\left(\frac{x}{2\sqrt{(\kappa t)}}\right)\right]$. To obtain this form, note that

$$\int_0^\infty e^{-\frac{1}{2}y^2} \cos vy \, dy = \sqrt{(\pi/2)} e^{-\frac{1}{2}v^2}$$

(cf. 9.19(2)), and integrate with respect to v over $[0, x/\sqrt{(2\kappa t)}]$. □

Notes (1) We used the Laplace transform on the t variable. If we had operated by $\mathcal{L}$ on x, we would have needed to know $u_x(0, t)$ (which we are not given) for the application of 9.4. A (good) alternative approach to this example is via the sine transform (see Supplementary exercise 9.9).

(2) If the boundary condition $u(0, t) = U$ is replaced by the more general condition $u(0, t) = g(t)$, the Convolution theorem enables one to obtain $u(x, t)$ in terms of an integral; cf. Example 9.24.

9.24 Example (Laplace's equation in a half-plane)
Suppose that $u(x, y)$ is defined and continuous on $\{(x, y) \in \mathbb{R}^2 : y \geqslant 0\}$ and satisfies

(i) $u_{xx} + u_{yy} = 0$;
(ii) $u(x, 0) = f(x)$ $(x \in \mathbb{R})$, where $f \in \mathscr{S}(\mathbb{R})$.
Show that under appropriate conditions on the behaviour of

$u(x, y)$ for large values of $r = (x^2 + y^2)^{\frac{1}{2}}$,

$$u(x, y) = \frac{y}{\pi} \int_{-\infty}^{\infty} \frac{f(t)}{(x-t)^2 + y^2} \, dt \qquad (y > 0).$$

Solution. We operate by the Fourier transform on the variable x and write

$$\hat{u}(s, y) = \int_{-\infty}^{\infty} u(x, y) e^{-isx} \, dx.$$

The partial differential equation (i) transforms into

$$\frac{d^2\hat{u}}{dy^2} = s^2 \hat{u};$$

here s is regarded as fixed. In deriving this we have used 9.15 and have assumed that derivatives of $\hat{u}$ with respect to y can be obtained by differentiation under the integral sign. The boundary condition (ii) transforms to $\hat{u}(s, 0) = \hat{f}(s)$, and, provided $u(x, y)$ decays sufficiently rapidly as $r \to \infty$, $\hat{u}(s, y) \to 0$ as $y \to \infty$, for each s. Then

$$\hat{u}(s, y) = \hat{f}(s) e^{-|s| y}.$$

By the Convolution theorem, 9.18, and Example 9.19(1),

$$u(x, y) = \frac{y}{\pi} \int_{-\infty}^{\infty} \frac{f(t)}{(x-t)^2 + y^2} \, dt. \qquad \square$$

Note We have here the Poisson integral for a half-plane. This is discussed further in 10.39(2).

9.25 Example (A Volterra integral equation)

To find a solution of

$$m(t) = (1 - e^{-\lambda t}) + \lambda \int_0^t m(t - x) e^{-\lambda x} \, dx \qquad (t \geq 0),$$

where λ is a positive constant.

Solution. Apply $\mathcal{L}$ and use the Convolution theorem. This gives

$$\bar{m}(p) = \frac{1}{p} - \frac{1}{\lambda + p} + \lambda \frac{\bar{m}(p)}{\lambda + p}.$$

Hence $\bar{m}(p) = \lambda/p^2$, so that $m(t) = \lambda t$ provides the required solution.
$$\square$$

Appendix: proofs of the Inversion and Convolution theorems

We allow Lebesgue integration the upper hand in this appendix, since it handles the results so much more cleanly than Riemann integration. For Riemann integrals our proofs are deficient on certain points of justification. We first prove the Inversion theorem for the Fourier transform, 9.17, and derive from it that for the Laplace transform, 9.8. For the reader's convenience we re-state the theorems.

The Inversion theorem for the Fourier transform

Suppose $f \in \mathcal{I}(\mathbb{R})$ is piecewise smooth. If

$$\hat{f}(s) = \int_{-\infty}^{\infty} f(x) e^{-isx} \, dx,$$

then

$$\tfrac{1}{2}[f(x+) + f(x-)] = \frac{1}{2\pi} \text{PV} \int_{-\infty}^{\infty} \hat{f}(s) e^{isx} \, ds \qquad (x \in \mathbb{R}).$$

Proof. Stage 1

$$\int_{-R}^{R} \hat{f}(s) e^{isx} \, ds = \int_{-R}^{R} \left(\int_{-\infty}^{\infty} f(u) e^{i(x-u)s} \, du \right) ds$$

$$= \int_{-\infty}^{\infty} \left(\int_{-R}^{R} f(u) e^{i(x-u)s} \, ds \right) du \qquad \text{(assuming the order of integration may be changed)}$$

$$= 2 \int_{-\infty}^{\infty} f(x+v) \frac{\sin Rv}{v} \, dv$$

$$= 2 \int_{0}^{\infty} [f(x+v) + f(x-v)] \frac{\sin Rv}{v} \, dv.$$

The interchange of order of integration is accomplished, for Lebesgue integrals, by the theorems of Fubini and Tonelli. A justification for Riemann integrals can be found in Apostol [1], 15–15.

Stage 2   The Riemann–Lebesgue lemma ([1], 15-6 or [8], p. 115) implies that for $\delta > 0$,

$$\lim_{R \to \infty} \int_\delta^\infty \frac{f(x+v)+f(x-v)}{v} \sin Rv \, dv = 0.$$

Stage 3 For $\delta > 0$,

$$\lim_{R \to \infty} \int_0^\delta \frac{\sin Rv}{v} \, dv = \lim_{R \to \infty} \int_0^{R\delta} \frac{\sin u}{u} \, du = \tfrac{1}{2}\pi \quad \text{(by 8.5).}$$

Stage 4 We are given that f is piecewise smooth. Provided we replace $f(x)$ by $f(x+)$ we can assume that f has a continuous derivative on $[x, x+\eta]$ for some $\eta > 0$. Let $\delta < \eta$ and $\delta \geqslant v > 0$. By the Mean-value theorem,

$$|f(x+v)-f(x+)| = v\,|f'(x+\theta v)| \text{ for some } \theta \ (0 < \theta < 1)$$
$$\leqslant Mv,$$

where M depends on η, but not on v or δ (since f' is continuous, and hence bounded, on $[x, x+\eta]$). Hence, for $\delta < \eta$,

$$\left| \int_0^\delta [f(x+v)-f(x+)] \frac{\sin Rv}{v} \, dv \right| \leqslant M\delta.$$

Similarly,

$$\int_0^\delta [f(x-v)-f(x-)] \frac{\sin Rv}{v} \, dv$$

can be made arbitrarily small by taking δ small enough.

Stage 5 Let $\varepsilon > 0$. Stage 1 shows that, for $\delta > 0$,

$$\frac{1}{2\pi} \int_{-R}^R \hat{f}(s) e^{isx} \, ds - \tfrac{1}{2}[f(x+)+f(x-)]$$

$$= \frac{1}{\pi} \int_\delta^\infty [f(x+v)+f(x-v)] \frac{\sin Rv}{v} \, dv$$

$$+ \frac{1}{\pi} \int_0^\delta [f(x+v)+f(x-v)-f(x+)-f(x-)] \frac{\sin Rv}{v} \, dv$$

$$+ \frac{1}{\pi} [f(x+)+f(x-)] \left(\int_0^\delta \frac{\sin Rv}{v} \, dv - \tfrac{1}{2}\pi \right)$$

$$=: I_1 + I_2 + I_3.$$

We can make $|I_2| < \varepsilon$ by taking δ sufficiently small (Stage 4). Fix such a δ. Then $|I_1| < \varepsilon$ if R is sufficiently large (Stage 2) and $|I_3| < \varepsilon$ if R is sufficiently large (Stage 3). □

Note The condition 'f piecewise smooth' was only used crucially in Stage 4. It could be replaced in the statement of the theorem by any condition sufficient to ensure $f(x\pm)$ exist for each x and

$$\lim_{\delta\to 0}\int_0^\delta [f(x\pm v)-f(x\pm)]\frac{\sin Rv}{v}\,dv=0.$$

The Inversion theorem for the Laplace transform

Let f be piecewise smooth on $[0,\infty)$ and suppose $\bar{f}(p)=\int_0^\infty f(t)e^{-pt}\,dt$ exists for Re $p>c\geqslant 0$. Then, if $t>0$,

$$\tfrac{1}{2}[f(t+)+f(t-)]=\frac{1}{2\pi i}\lim_{R\to\infty}\int_{\sigma-iR}^{\sigma+iR}\bar{f}(p)e^{pt}\,dp\qquad (\sigma>c).$$

Proof. For p on the line of integration, $p=\sigma+iy$ and

$$\bar{f}(p)=\int_0^\infty [e^{-\sigma t}f(t)]e^{-iyt}\,dt.$$

Apply the Fourier inversion theorem to $e^{-\sigma t}f(t)\chi_{[0,\infty)}(t)$, which is piecewise smooth and in $\mathscr{S}(\mathbb{R})$, to get, for $\sigma>c$ and $t>0$,

$$\tfrac{1}{2}e^{-\sigma t}[f(t+)+f(t-)]=\frac{1}{2\pi}\lim_{R\to\infty}\int_{-R}^R \bar{f}(\sigma+iy)e^{iyt}\,dy.$$

We obtain the required inversion formula by writing p for $\sigma+iy$.
□

Convolutions

We included convolutions in our discussion of transforms as a valuable adjunct to our other inversion techniques. We should stress that they are also of importance in their own right: they arise naturally in probability theory, for example. We begin by stating the Lebesgue version of the Convolution theorem for the Fourier transform, with greater precision than in 9.18.

The Convolution theorem for the Fourier transform

Suppose f and g belong to $L^1(\mathbb{R})$. Then $h(x)=\int_{-\infty}^\infty f(y)g(x-y)\,dy$ exists almost everywhere and defines a function $h\in L^1(\mathbb{R})$ such that $\hat{h}(s)=\hat{f}(s)\hat{g}(s)$ $(s\in\mathbb{R})$.

Proof. We apply Tonelli's theorem and Fubini's theorem (Weir [8], p. 123 and p. 83). The function $f(y)g(x-y)$ can be shown to be measurable as a function of (x, y) on $\mathbb{R}^2$. (We omit the technical proof because measurability is not in doubt in any of our applications of the Convolution theorem.) Now, for $s \in \mathbb{R}$,

$$\iint |f(y)g(x-y)e^{-isx}|\, dx\, dy = \int |f(y)| \left(\int |g(x-y)|\, dx \right) dy$$

$$= \int |f(y)|\, dy \int |g(v)|\, dv$$

(putting $v = x - y$ in the inner integral (with y fixed)) and this is finite since f and g are both in $L^1(\mathbb{R})$. Tonelli's theorem implies that $f(y)g(x-y)e^{-isx}$ is integrable on $\mathbb{R}^2$. Take $s = 0$ (we need the general case later). Fubini's theorem implies that

$$h(x) = \int f(y)g(x-y)\, dy$$

exists almost everywhere and that h belongs to $L^1(\mathbb{R})$.

We now establish the convolution formula. We have

$$\hat{h}(s) = \int h(x)e^{-isx}\, dx$$

$$= \iint f(y)g(x-y)e^{-isx}\, dy\, dx$$

$$= \iint f(y)g(x-y)e^{-isx}\, dx\, dy \quad \text{(provided the order of integration may be changed)}$$

$$= \int f(y)e^{-isy} \left(\int g(v)e^{-isv}\, dv \right) dy \quad \text{(putting } v = x - y \text{ in the inner integral (with } y \text{ fixed))}$$

$$= \hat{f}(s)\hat{g}(s).$$

To justify the interchange of order of integration we merely need to invoke Fubini's theorem, since we showed above that $f(y)$ $g(x-y)e^{-isx}$ is integrable on $\mathbb{R}^2$, for each s.

We shall not prove in detail the corresponding result for Riemann integrals. However we note that our definition of $\mathscr{I}(\mathbb{R})$ in 9.1 is such that, if f and g belong to $\mathscr{I}(\mathbb{R})$, then the convolution integral defining h makes sense everywhere; see Apostol [1], 15–18. The formula $\hat{h}(s) = \hat{f}(s)\hat{g}(s)$ is derived in just the same way as above, except for the way the interchange of the order of the integrals is treated; see [1], 5–19.

To avoid repetition, we derive the Convolution theorem for Laplace transforms from that for Fourier transforms. The reader is however urged to write out a direct proof as an exercise. The recommended method, given f and g as in the theorem, is to work with $f(x)g(t-x)e^{-pt}k(x, t)$, where $k(x, t) = 1$ if $0 \leqslant x \leqslant t$ and 0 otherwise. The function k conveniently builds the variable limits of integration into the integrand.

The Convolution theorem for the Laplace transform

Suppose f and g are functions on $[0, \infty)$ such that $\bar{f}(p)$ and $\bar{g}(p)$ exist for Re $p > c$. Define h by

$$h(t) = \int_0^t f(s)g(t-s)\,\mathrm{d}s \quad (\text{for } t \geqslant 0).$$

Then $\bar{h}(p)$ exists for Re $p > c$ and $\bar{h}(p) = \bar{f}(p)\bar{g}(p)$.

Proof. Let $p = u + \mathrm{i}s$ $(u > c)$ and define, for $t \in [0, \infty)$,

$$f_1(t) = f(t)e^{-ut}\chi_{[0,\infty)}(t) \text{ and } g_1(t) = g(t)e^{-ut}\chi_{[0,\infty)}(t).$$

Then f_1 and g_1 are both in $\mathscr{I}(\mathbb{R})$ and their Fourier convolution is h_1, where

$$h_1(t) = e^{-ut}\int_{-\infty}^{\infty} f(x)g(t-x)\chi_{[0,\infty)}(x)\chi_{[0,\infty)}(t-x)\,\mathrm{d}x$$

$$= e^{-ut}\chi_{[0,\infty)}(t)\int_0^t f(x)g(t-x)\,\mathrm{d}x.$$

Hence $h_1(t) = e^{-ut}h(t)$ $(t \geqslant 0)$ and we get

$$\bar{h}(p) = \hat{h}_1(s) = \hat{f}_1(s)\hat{g}_1(s) = \bar{f}(p)\bar{g}(p). \qquad \square$$

Exercises

1. Verify the entries in Table 9.1. Use 9.9 to confirm that the Inversion theorem, 9.8, gives the expected results for the functions listed.

2. Find the Laplace transform of (i) $t(t^2 - 1)$, (ii) $t(\cos t)e^{-t}$, (iii) $\cosh t \cos t$, (iv) $\chi_{[0, T]}(t)$ $(T > 0)$.

3. The Laguerre polynomials are defined by

$$L_n(t) := \frac{e^t}{n!}\left(\frac{\mathrm{d}}{\mathrm{d}t}\right)^n(t^n e^{-t}) \qquad (n = 0, 1, 2, \ldots).$$

Prove that $\mathscr{L}[L_n(t)] = (p-1)^n/p^{n+1}$, and deduce that

$$t\frac{\mathrm{d}^2}{\mathrm{d}t^2}L_n + (1-t)\frac{\mathrm{d}}{\mathrm{d}t}L_n + nL_n = 0.$$

4. Find the inverse Laplace transform of (i) $(p(p+1)(p+2))^{-1}$, (ii) $(p^2-1)^{-2}$, (iii) $6(p^4+10p^2+9)^{-1}$, (iv) $2p/(p^4+1)$.

(i) Use 9.19(2) to show that

$$\int_0^\infty e^{-\alpha u^2} \cos \beta u \, du = \frac{1}{2}\sqrt{\frac{\pi}{\alpha}} e^{-\beta^2/(4\alpha)} \quad (\alpha > 0).$$

(ii) Let $p^{\frac{1}{2}}$ be given by $p^{\frac{1}{2}} = |p|^{\frac{1}{2}} e^{i\theta/2}$, where θ lies between $-\pi$ and π. By integrating round the 'keyhole' contour in Figure 9.4, show that

$$\mathcal{L}^{-1}[e^{-2p^{\frac{1}{2}}}/p^{\frac{1}{2}}] = \frac{1}{\pi} \int_0^\infty \frac{\cos(2\sqrt{x})e^{-xt}}{\sqrt{x}} dx.$$

Deduce that

$$\mathcal{L}[e^{-1/t}/\sqrt{t}] = \sqrt{\pi}e^{-2p^{\frac{1}{2}}}/p^{\frac{1}{2}}.$$

5. Find the inverse Fourier transform of

(i) $1/(is+1)$, (ii) $(1-s^2)(1+s^2)^{-2}$, (iii) $se^{-s}\chi_{[0,\infty)} - se^s\chi_{(-\infty,0)}$, (iv) $s^{-1}\sin s$.

6. Suppose that f satisfies the hypotheses of the Fourier inversion theorem, 9.17, and that $f(x) = f(-x)$ for all x. Show that

$$\frac{1}{2}[f(x+) + f(x-)] = \frac{2}{\pi}\int_0^\infty \cos vx \left\{\int_0^\infty f(y)\cos yv \, dy\right\} dv.$$

Hence evaluate, for $a \in \mathbb{R}$,

(i) $\displaystyle\int_0^\infty e^{-v^2}\cos 2av \, dv$, (ii) $\displaystyle\int_0^\infty \frac{\sin av \cos av}{v} dv$.

7. Use the Laplace transform to solve, for $t>0$, the equation

$$\frac{d^2y}{dt^2} - 6\frac{dy}{dt} + 13y = 0,$$

subject to $y(0) = y'(0) = 1$.

8. Solve, for $t \geq 0$, the simultaneous equations

$$\frac{d^2x}{dt^2} + 2n\frac{dx}{dt} + n^2x = 0, \qquad \frac{dy}{dt} + 2n\frac{d^2y}{dt^2} + n^2y = \mu\frac{dx}{dt},$$

where μ is a constant, $x(0) = x'(0) = 1$, and $y(0) = y'(0) = 0$.

9. Suppose that x, y, and z are functions on $[0, \infty)$ such that

$$\frac{dx}{dt} = bz - cy, \qquad \frac{dy}{dt} = cx - az, \qquad \frac{dz}{dt} = ay - bx,$$

where a, b, and c are constants. Show that, if $x(0) = 1$ and $y(0) = z(0) = 0$,

$$x(t) = [a^2 + (b^2+c^2)\cos \omega t]/\omega^2, \quad \text{where} \quad \omega^2 = a^2+b^2+c^2.$$

10. Solve, for $t > 0$, the integral equation

$$y(t) = 1 + \int_0^t x e^{-x} y(t-x) \, dx.$$

11. The functions $u_0, u_1, u_2, \ldots$ are related by the equations

$$u_n'(t) = u_{n-1}(t) - u_n(t) \qquad (n \geq 1, \, t \geq 0).$$

Use the Laplace transform to prove that

$$u_n(t) = \int_0^t \phi_{n-1}(t-x) u_0(x) \, dx + \sum_{r=1}^n \phi_{n-r}(t) u_r(0) \qquad (n \geq 1, \, t \geq 0),$$

where the functions $\phi_1, \phi_2, \ldots$ are to be determined.

12. Use the Laplace transform to find $u(x, t)$ satisfying, for $x > 0$ and $t > 0$,

$$\frac{\partial^2 u}{\partial t^2} - \frac{\partial^2 u}{\partial x^2} = t e^{-x},$$

$$u(x, 0) = 0, \qquad u_t(x, 0) = x, \qquad u(0, t) = 1 - e^{-t}.$$

10 Conformal mapping and harmonic functions

The first part of the chapter concerns angle-preserving mappings between regions in the complex plane. Such mappings are of intrinsic geometric interest and are important in advanced complex analysis. They are also worth studying because of their usefulness in solving problems concerning harmonic functions in $\mathbb{R}^2$ (in particular, certain two-dimensional problems in fluid dynamics). The final section reveals the striking parallels between the theory of holomorphic functions and that of harmonic functions.

We shall adopt a more cavalier attitude to the argument of a complex number than previously. This is legitimate since we are concerned with the determination of argument 'a point at a time' and not with the variation of the argument of a moving point. We accordingly write $\arg z$ to denote any choice from $[\arg z] = \{\theta : z = |z|\, e^{i\theta}\}$. The price to pay is that some equations only hold modulo an integer multiple of 2π.

It will sometimes be convenient to work in the extended plane $\tilde{\mathbb{C}}$ introduced in 6.13. We adopt the following conventions:

$$a \pm \infty = \pm\infty + a = \infty, \qquad a/\infty = 0 \quad \text{for all } a \in \mathbb{C},$$

$$a \times \infty = \infty \times a = \infty, \qquad a/0 = \infty \quad \text{for all } a \in \mathbb{C}\backslash\{0\},$$

$$\infty + \infty = \infty \times \infty = \bar{\infty} = \infty.$$

Circles and lines revisited

Any circle is given by an equation $|z - a| = r$. However our repertoire of techniques for solving mapping problems involving circles will be greatly enlarged by having another form of equation available.

We work in $\tilde{\mathbb{C}}$, and adjoin ∞ to any line in $\mathbb{C}$. Then circles and straight lines in $\tilde{\mathbb{C}}$ correspond to circles on the Riemann sphere Σ, with straight lines associated with circles passing through the north pole. The distinction between circles and lines is thus rather an artificial one, and we introduce the generic name *circline* to cover both.

10.1 Definition

Points α and β are *inverse with respect to the circle* $|z-a|=r$ if $(\alpha-a)\overline{(\beta-a)}=r^2$ (we include the pair $\alpha=a$, $\beta=\infty$); see Fig. 10.1 below. Note that α, β, and a are collinear. Points α and β are *inverse with respect to a straight line* ℓ if β is the reflection of α in ℓ.

10.2 Theorem (Inverse point representation of circlines)

Let α and β belong to $\mathbb{C}$, $\alpha\neq\beta$. Then for any $\lambda>0$, the equation $\left|\dfrac{z-\alpha}{z-\beta}\right|=\lambda$ represents a circline with inverse points α and β, and every such circline can be so represented.

Proof. Consider the equation $\left|\dfrac{z-\alpha}{z-\beta}\right|=\lambda$. Denote α by A, β by B, and the variable point z by P. If $\lambda=1$, the locus of P is the perpendicular bisector of AB; it has α and β as inverse points.

Now assume $\lambda\neq1$. The equation gives the locus of points P for which the ratio $AP:PB$ has the constant value λ. The locus is a circle (known as the circle of Apollonius). This can be proved geometrically, but it is simpler to switch to cartesian coordinates (a strategy usually to be avoided in complex analysis). Put $\alpha=\alpha_1+i\alpha_2$, $\beta=\beta_1+i\beta_2$, and $z=x+iy$. The equation becomes

$$\left(x-\frac{\alpha_1-\lambda^2\beta_1}{1-\lambda^2}\right)^2+\left(y-\frac{\alpha_2-\lambda^2\beta_2}{1-\lambda^2}\right)^2=K,$$

where K is a constant, and this certainly represents a circle. Define z_1 and z_2 by

$$\alpha-z_1=\lambda(z_1-\beta)\quad\text{and}\quad\alpha-z_2=\lambda(\beta-z_2).$$

These points lie on the circle and are collinear with α and β, so they are endpoints of a diameter (see Fig. 10.1). The circle has centre $a=\frac{1}{2}(z_1+z_2)$ and radius $r=\frac{1}{2}|z_1-z_2|$. Adding and subtracting the equations above gives

$$\alpha-a=\tfrac{1}{2}\lambda(z_1-z_2)\quad\text{and}\quad 2\lambda(\beta-a)=(z_1-z_2).$$

Hence $(\alpha-a)\overline{(\beta-a)}=\tfrac{1}{4}(z_1-z_2)\overline{(z_1-z_2)}=r^2$, which proves that α and β are inverse points with respect to the circle.

Conversely, it is clear that the equation of any line can be written as $|z-\alpha|=|z-\beta|$ for some α and β. Also, given any circle $|z-a|=r$, we can choose $\alpha\neq a$ and then $\beta=a+r^2\overline{(\alpha-a)}^{-1}$ to make α and β inverse points. Take any point z_0 on the circle and

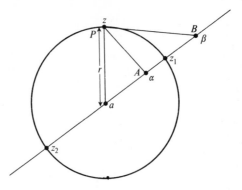

Fig. 10.1

let $\lambda = |z_0 - \alpha| / |z_0 - \beta|$. Then the circle $|(z - \alpha)/(z - \beta)| = \lambda$ coincides with the given one. $\qquad\square$

Note A circline is uniquely determined by a pair of inverse points and a point on the circline.

10.3 The unit circle ($|z| = 1$)

Any points α and $1/\bar{\alpha}$ in $\mathbb{C}$ are inverse with respect to the unit circle, and the point 1 lies on the circle. Hence, by Theorem 10.2, the equation can be written

$$\left| \frac{z - \alpha}{z - 1/\bar{\alpha}} \right| = \left| \frac{1 - \alpha}{1 - 1/\bar{\alpha}} \right|, \quad \text{that is,}$$

$$\left| \frac{z - \alpha}{\bar{\alpha}z - 1} \right| = 1 \quad (\text{since } |\alpha - 1| = |\bar{\alpha} - 1|).$$

10.4 Representation of circular arcs

If P is a variable point on a circular arc with endpoints A and B, then $\mu = \angle APB$ is constant. From Fig. 10.2, $\mu = \theta - \phi$, where $\arg(z - \alpha) = \theta$ and $\arg(z - \beta) = \phi$. Hence the arc APB has equation

$$\arg(z - \alpha) - \arg(z - \beta) = \mu \quad (\text{mod } 2\pi),$$

that is,

$$\arg\left(\frac{z - \alpha}{z - \beta} \right) = \mu \quad (\text{mod } 2\pi) \qquad (z \neq \alpha, \beta).$$

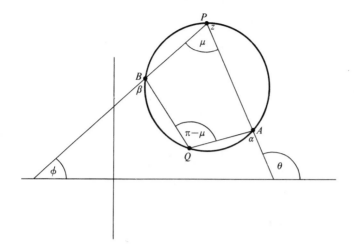

Fig. 10.2

Similarly the equation of the arc AQB is (note the signs!)

$$\arg\left(\frac{z-\alpha}{z-\beta}\right) = -(\pi-\mu) \quad (\text{mod } 2\pi) \qquad (z \neq \alpha, \beta).$$

10.5 Coaxal circles

For fixed α and β we have, as λ and μ vary, two families of circles:
(1) $C_1(\alpha, \beta)$: circles

$$\left|\frac{z-\alpha}{z-\beta}\right| = \lambda,$$

having α and β as inverse points, and
(2) $C_2(\alpha, \beta)$: circles

$$\arg\left(\frac{z-\alpha}{z-\beta}\right) = \begin{cases} \mu \\ -(\pi-\mu) \end{cases} \quad (\text{mod } 2\pi),$$

through α and β.

Each of the families is said to form a *coaxal system*. These systems of circles have interesting geometric properties. It can be shown, for example, that any member of $C_1(\alpha, \beta)$ cuts any member of $C_2(\alpha, \beta)$ orthogonally.

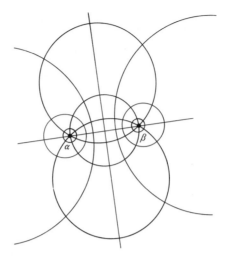

Fig. 10.3

Conformal mapping

We now investigate how holomorphic functions behave when considered as geometric mappings.

10.6 Theorem

Suppose f is holomorphic in an open set G, $\zeta \in G$, and $f'(\zeta) \neq 0$. Then, in the sense defined below, f preserves angles between paths in G meeting at ζ.

Proof. Let γ_1 and γ_2 be paths lying in G, both with parameter interval $[0,1]$, having common endpoint $\zeta = \gamma_1(0) = \gamma_2(0)$. We suppose that, for $k = 1$ and 2, $\gamma_k'(0) \neq 0$, so that γ_k has a well-defined tangent at ζ given by $\gamma_k(t) = \zeta + \gamma_k'(0)t$ $(t \geq 0)$ and making an angle $\arg \gamma_k'(0)$ with the real axis. The angle between γ_1 and γ_2 is then (by definition) $\lambda = \arg \gamma_1'(0) - \arg \gamma_2'(0)$.

The paths γ_1 and γ_2 are mapped by f to paths $f \circ \gamma_1$ and $f \circ \gamma_2$ and these meet at $f(\zeta)$ at an angle $\Lambda = \arg(f \circ \gamma_1)'(0) - \arg(f \circ \gamma_2)'(0)$. The assertion of the theorem is that $\Lambda = \lambda \pmod{2\pi}$. By the chain rule,

$$\frac{(f \circ \gamma_1)'(0)}{(f \circ \gamma_2)'(0)} = \frac{f'(\zeta)\gamma_1'(0)}{f'(\zeta)\gamma_2'(0)} = \frac{\gamma_1'(0)}{\gamma_2'(0)},$$

from which the result follows; see 2.18. □

10.7 Definition

A mapping f is *conformal in an open set* G if $f \in H(G)$ and $f'(z) \neq 0$ for any $z \in G$; f is *conformal at a point* ζ if it is conformal in some disc $D(\zeta; r)$.

The proof of Theorem 10.6 shows that conformal mappings preserve both the magnitude and sense of angles. If f is differentiable at ζ but $f'(\zeta) = 0$, f does not preserve angles at ζ. Consider, for example, $f(z) = z^2$, which doubles angles at 0.

10.8 Construction of conformal mappings: preliminary remarks

(1) Suppose we require a conformal map f from the open upper half-plane $\Pi = \{z : \operatorname{Im} z > 0\}$ onto the open unit disc $D(0; 1)$. It is unlikely to be helpful to bring $\operatorname{Im} z$ (a non-holomorphic function) directly into the definition of f (which must be holomorphic). Recall however that $\Pi = \{z : |z - i| < |z + i|\}$ (the set of points closer to i than to $-i$). It ought now to be clear that we should take $f(z) = (z - i)/(z + i)$ for then

$$z \in \Pi \Leftrightarrow |f(z)| < 1 \Leftrightarrow f(z) \in D(0; 1).$$

Also f is conformal in Π since f is holomorphic there, with $f'(z) = 2i(z + i)^{-2} \neq 0$. This simple example shows that success in constructing a conformal mapping from one region to another may depend on a judicious choice of descriptions for the regions. Later examples will reinforce this point.

(2) The composition of conformal mappings is conformal, by the chain rule. Hence we can hope to build up a conformal mapping from one region to another by taking a finite sequence of 'elementary' conformal mappings. For example, a typical sequence of mappings from a lozenge (bounded by circular arcs) to $D(0; 1)$ might be as shown in Fig. 10.4. Thus an aid to successful map-building is familiarity with standard mappings. These include the

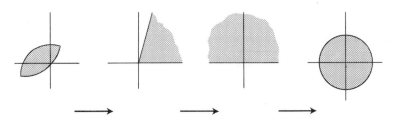

Fig. 10.4

Möbius transformations, exponentials, and powers discussed below.

Möbius transformations

10.9 Definition

A Möbius transformation is a mapping of the form

$$f : z \mapsto \frac{az + b}{cz + d} \quad (a, b, c, d \in \mathbb{C}, \, ad - bc \neq 0).$$

(The excluded case $ad - bc = 0$ produces a constant mapping.)

The Möbius transformation f is best viewed as a mapping of $\tilde{\mathbb{C}}$ to itself, with (by definition) $f(-d/c) = \infty$, $f(\infty) = a/c$. The map $f : \tilde{\mathbb{C}} \to \tilde{\mathbb{C}}$ is one-to-one and onto, with inverse

$$f^{-1} : w \mapsto \frac{dw - b}{a - cw}$$

also a Möbius transformation. It is easily checked that the Möbius transformations form a group under the operation of composition of maps. A general Möbius transformation can be built up from mappings of the following types:

$z \mapsto z e^{i\phi}$ (ϕ real) (anticlockwise rotation through ϕ),

$z \mapsto Rz$ ($R > 0$) (stretching by a factor of R),

$z \mapsto z + a$ ($a \in \mathbb{C}$) (translation by a),

$z \mapsto 1/z$ (inversion).

Suppose $f(z) = (az + b)/(cz + d)$. Then $f'(z) = (ad - bc)/(cz + d)^2$, which shows that f is conformal on $\mathbb{C} \setminus \{-d/c\}$. Theorem 10.10 shows just how much room for manoeuvre we have in constructing Möbius transformations.

10.10 Theorem

Suppose each of $\{z_1, z_2, z_3\}$ and $\{w_1, w_2, w_3\}$ is a triple of distinct points in $\tilde{\mathbb{C}}$. Then there exists a unique Möbius transformation f such that $f(z_k) = w_k$ ($k = 1, 2, 3$), given by $f : z \mapsto w = f(z)$, where

$$\left(\frac{w - w_1}{w - w_3} \right) \left(\frac{w_2 - w_3}{w_2 - w_1} \right) = \left(\frac{z - z_1}{z - z_3} \right) \left(\frac{z_2 - z_3}{z_2 - z_1} \right).$$

Proof. The map

$$g : z \mapsto \left(\frac{z - z_1}{z - z_3}\right)\left(\frac{z_2 - z_3}{z_2 - z_1}\right)$$

takes z_1, z_2, and z_3 to 0, 1, and ∞, respectively. Construct h in the same way as g, to map w_1, w_2, and w_3 to $0, 1$, and ∞. Then $f = h^{-1} \circ g$ is the map in the statement of the theorem and $f(z_k) = w_k \, (k = 1, 2, 3)$.

For uniqueness it is enough to show that the only Möbius transformation $f : z \mapsto (az + b)/(cz + d)$ fixing 0, 1, and ∞ is the identity map. The conditions $f(0) = 0$, $f(\infty) = \infty$, and $f(1) = 1$ force in turn $b = 0, c = 0$, and $a = d$, so that $f(z) = z$ for all z. □

One can show that Möbius transformations map circlines to circlines by using any of the standard representations of circles and lines. The proof below shows more: the preservation of inverse points.

10.11 Theorem

Let S be a circline with inverse points α and β ($\alpha, \beta \in \mathbb{C}$, $\alpha \neq \beta$) and let f be a Möbius transformation. Then f maps S to a circline with inverse points $f(\alpha)$ and $f(\beta)$.

Proof. We use Theorem 10.2 to write the equation of S in the form $|(z - \alpha)/(z - \beta)| = \lambda$. Suppose $w = f(z) = (az + b)/(cz + d)$, so $z = (dw - b)/(a - cw)$. Then

$$\left| \frac{(\alpha c + d)w - (\alpha a + b)}{(\beta c + d)w - (\beta a + b)} \right| = \lambda.$$

We may rewrite this as

(i) $\left| \dfrac{w - f(\alpha)}{w - f(\beta)} \right| = \lambda \left| \dfrac{\beta c + d}{\alpha c + d} \right|$, if $\alpha c + d \neq 0$ and $\beta c + d \neq 0$, or

(ii) $|w - f(\alpha)| = \lambda \left| \dfrac{\beta a + b}{\alpha c + d} \right|$, if $\alpha c + d \neq 0$ and $\beta c + d = 0$, or

(iii) $|w - f(\beta)| = \dfrac{1}{\lambda} \left| \dfrac{\alpha a + b}{\beta c + d} \right|$, if $\alpha c + d = 0$ and $\beta c + d \neq 0$.

(Note that $\alpha c + d$ and $\beta c + d$ cannot both be zero.) In each case the image of S is a circline with $f(\alpha)$ and $f(\beta)$ as inverse points; in cases (ii) and (iii) the images of the original inverse points are the centre of the circle and the point ∞. □

10.12 Theorem

There exists a Möbius transformation mapping any given circline to any other given circline.

Proof. The result follows immediately from Theorems 10.10 and 10.11, since a circline is uniquely determined by three points (these may be taken to be a pair of inverse points and one point on the circline, or three points on it). □

10.13 Examples

(1) To find the images under $f: z \mapsto w = (z-1)^{-1}$ of (a) the real axis, (b) the imaginary axis, (c) the circle centre 0, radius r.

Solution (a) The real axis is mapped to the circline through $f(0) = -1$, $f(1) = \infty$, $f(\infty) = 0$, viz. the real axis. (Alternatively one may use the fact that the real axis has equation $z = \bar{z}$.)
(b) The imaginary axis has equation $|z-1| = |z+1|$ and is mapped to the circline with equation $|2w+1| = 1$, which is the circle with centre $-\frac{1}{2}$ and radius $\frac{1}{2}$.
(c) The required image has equation $|w+1| = r|w|$. If $r = 1$, it is the line Re $w = -\frac{1}{2}$. If $r \neq 1$, it is a circle with -1 and 0 as inverse points. It has the points $(\pm r - 1)^{-1}$ as the ends of a diameter (see Fig. 10.1), and hence has centre $(r^2 - 1)^{-1}$ and radius $r|1-r^2|^{-1}$. □
(2) To find all Möbius transformations mapping the unit circle T to itself and mapping α to 0.

Solution. Let f satisfy the required conditions and map 1 to $e^{i\phi} \in T$. The points α and $1/\bar{\alpha}$ are inverse points with respect to T and so are mapped by f to 0 and ∞ (by 10.11). The unique Möbius transformation taking α, 1, and $0/\bar{\alpha}$ to 1, $e^{i\phi}$, and ∞, respectively, is given by Theorem 10.10 to be $z \mapsto w$ where

$$\frac{w}{e^{i\phi}} = \left(\frac{z-\alpha}{z-1/\bar{\alpha}}\right)\left(\frac{1-1/\bar{\alpha}}{1-\alpha}\right) = \left(\frac{z-\alpha}{\bar{\alpha}z-1}\right)\left(\frac{\bar{\alpha}-1}{1-\alpha}\right).$$

But $|1-\alpha| = |1-\bar{\alpha}|$, so, for some real number ψ,

$$w = e^{i\psi}\left(\frac{z-\alpha}{\bar{\alpha}z-1}\right).$$

□

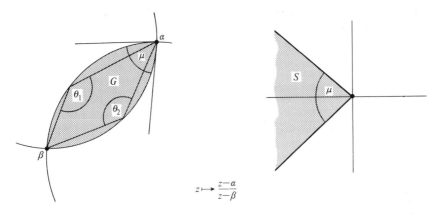

Fig. 10.5

10.14 Mappings of regions bounded by circular arcs

A Möbius transformation can be used, *inter alia*, to map a lozenge-shaped region bounded by circular arcs onto a sector. Let G be as shown in Fig. 10.5. By 10.4,

$$G = \left\{ z : \theta_1 < \arg\left(\frac{z-\alpha}{z-\beta}\right) < 2\pi - \theta_2 \right\}.$$

The Möbius transformation $f : z \mapsto (z-\alpha)/(z-\beta)$ maps G conformally onto the sector

$$S = \{w : \theta_1 < \arg w < 2\pi - \theta_2\}.$$

Note the effect of mapping β to ∞; circular arcs ending at β transform to half-lines. Observe also that, because f is conformal at α, the angle subtended by S at 0 equals the angle μ between the bounding arcs of G. Note that the image sector is determined completely by the image of a single point other than α or β. By taking the map $z \mapsto k(z - \alpha)/(z - \beta)$, with $k = e^{i\varphi}$ ($\varphi \in \mathbb{R}$) chosen suitably, the image sector can be swung round into any desired position.

This argument deals with mappings of regions bounded by two members of a coaxal system $C_2(\alpha, \beta)$ as defined in 10.5. Exercise 10.8 concerns the mapping of a region bounded by two non-concentric circles, which form members of some coaxal system $C_1(\alpha, \beta)$.

Other mappings: powers, exponentials, and the Joukowski transformation

10.15 Powers

The map $z \mapsto z^n$ $(n = 2, 3, \ldots)$ is conformal except at 0, where angles between paths are magnified by a factor of n. With non-integer powers we have to contend with a multifunction. Suppose, for definiteness, that we define, for $\alpha \in \mathbb{R}$, $z^\alpha = |z|^\alpha e^{i\theta\alpha}$ $(z = |z| e^{i\theta}, -\pi < \theta \leqslant \pi)$. Then $z \mapsto z^\alpha$ is conformal in the plane cut along $(-\infty, 0]$. Particularly useful is $z \mapsto z^{\pi/\beta}$; for $0 < \beta < \pi$, this takes the sector $\{z : 0 < \arg z < \beta\}$ conformally onto the open upper half-plane (see Fig. 10.6).

10.16 Exponentials

Let $f : z = x + iy \mapsto e^z = w = R e^{i\phi}$; this map is conformal in $\mathbb{C}$. Since $R = e^x$ and $\phi = g \pmod{2\pi}$, f maps

> a line $x = a$ to a circle $|w| = e^a$, and
>
> a line $y = c$ to a half-line $\arg w = c$.

Hence f takes the vertical strip $\{z : a < \operatorname{Re} z < b\}$ to the annulus $\{w : e^a < |w| < e^b\}$ and the horizontal strip $\{z : c < \operatorname{Im} z < d\}$ to the sector $\{w : c < \arg w < d\}$; see Fig. 10.7. In reverse, a logarithm will map sectors to strips, but we need to select a holomorphic branch and to work in the appropriate cut plane.

10.17 The Joukowski transformation

Möbius transformations, powers and exponentials have the property of mapping certain families of circles and straight lines to

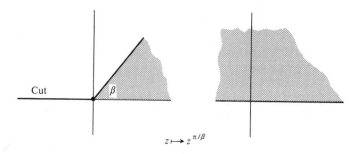

$$z \mapsto z^{\pi/\beta}$$

Fig. 10.6

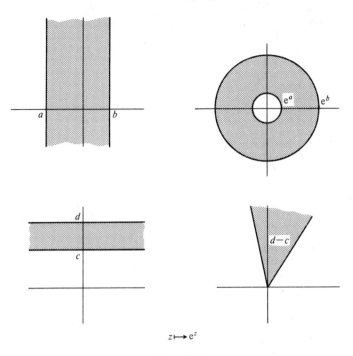

$$z \mapsto e^z$$

Fig. 10.7

similar families. As an example of a mapping of a different type we consider the simplest form of Joukowski transformation,

$$z \mapsto w = \tfrac{1}{2}(z + z^{-1}).$$

This is given equivalently by

$$\frac{w+1}{w-1} = \left(\frac{z+1}{z-1}\right)^2.$$

It is holomorphic except at 0 and ∞, and conformal except at ± 1, where angles are doubled.

Suppose $w = u + iv$ is the image of $z = re^{i\theta}$, so that

$$u = \tfrac{1}{2}(r + r^{-1})\cos \theta, \quad v = \tfrac{1}{2}(r - r^{-1})\sin \theta.$$

Then the image of the circle $|z| = \rho$ is the ellipse

$$\frac{u^2}{\tfrac{1}{4}(\rho + \rho^{-1})^2} + \frac{v^2}{\tfrac{1}{4}(\rho - \rho^{-1})^2} = 1,$$

while the image of the half-line $\arg z = \mu$ is

$$\frac{u^2}{\cos^2\mu} - \frac{v^2}{\sin^2\mu} = 1,$$

which is a hyperbola.

Examples on building conformal mappings

This section gives examples to show how we can combine the mappings introduced above to map an assortment of regions onto simpler ones such as discs and half-planes.

10.18 Example

To find a conformal mapping of the semicircular region $G = \{z : \operatorname{Im} z > 0 , |z| < 1\}$ onto $D(0; 1)$.

Solution.

Stage 1 By 10.14, we can express G as

$$\left\{z : \tfrac{1}{2}\pi < \arg\left(\frac{z-1}{z+1}\right) < \pi\right\}$$

(note that G is bounded by arcs through ± 1 subtending angles $\tfrac{1}{2}\pi$ and 0). Define $g(z) = (z-1)/(z+1) = \zeta$. Under g, G is mapped conformally to the quadrant $Q = \{\zeta : \tfrac{1}{2}\pi < \arg \zeta < \pi\}$.

Stage 2 Let $\tau = \zeta^2$. Under $\zeta \mapsto \tau$, Q is mapped conformally to the open lower half-plane $\{\tau : \pi < \arg \tau < 2\pi\}$. (Here the non-conformality of $\zeta \mapsto \zeta^2$ at $\zeta = 0$ ($\notin Q$) works to our advantage.)

Stage 3 The open lower half-plane is $\{\tau : |\tau + i| < |\tau - i|\}$ and so is mapped to $\{w : |w| < 1\}$ under $\tau \mapsto (\tau + i)/(\tau - i) = w$.
    A suitable map f can now be seen to be

$$f : z \mapsto w = \frac{(z-1)^2 + i(z+1)^2}{(z-1)^2 - i(z+1)^2} = i\left(\frac{z^2 + 2iz + 1}{z^2 - 2iz + 1}\right);$$

this maps G onto $D(0; 1)$ by construction and, as the composite of conformal maps, is conformal. ☐

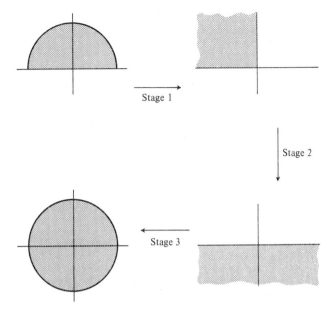

Fig. 10.8

10.19 Example

To find a conformal mapping of the semi-infinite strip $H = \{z : \operatorname{Im} z > 0, 0 < \operatorname{Re} z < \pi\}$ onto a half-plane.

Solution.

Stage 1 Put $\zeta = e^{iz}$. Then $|\zeta| = e^{-\operatorname{Im} z}$ and $\arg \zeta = \operatorname{Re} z \pmod{2\pi}$. So H is mapped by $z \mapsto \zeta$ conformally onto

$$G = \{\zeta : 0 < |\zeta| < 1, 0 < \arg \zeta < \pi\}.$$

Stage 2 We could now proceed as in Stages 1 and 2 of Example 10.18. More directly, we can use the Joukowski transformation $\zeta \mapsto w = \frac{1}{2}(\zeta + \zeta^{-1})$, which is conformal in G. The region G is the union, over $0 < r < 1$, of the semicircular arcs $\zeta = r e^{i\theta}$ $(0 < \theta < \pi)$; see Fig. 10.9. Each of these arcs is mapped to an elliptic arc

$$\frac{u^2}{\frac{1}{4}(r + r^{-1})^2} + \frac{v^2}{\frac{1}{4}(r - r^{-1})^2} = 1, \quad v < 0 \quad (\zeta = u + iv).$$

The union of these images covers the open lower half-plane, onto which H is thus mapped by the composite transformation

$$z \mapsto w = \frac{1}{2}(e^{iz} + e^{-iz}) = \cos z.$$

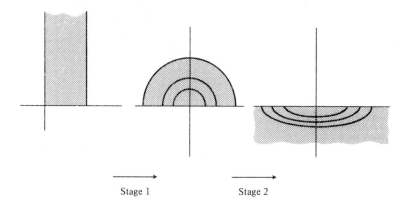

Stage 1   Stage 2

Fig. 10.9

10.20 Example

To find a conformal mapping of the region G exterior to both the circles $|z \pm 1| = \sqrt{2}$ onto the region $\hat{G}$ exterior to the unit circle.

Solution.

Stage 1 The region G is bounded by circular arcs meeting orthogonally at $\pm i$ (see Fig. 10.10). Take

$$g : z \mapsto \frac{z-i}{z+i} = \zeta;$$

g is conformal except at $-i \notin G$. The boundary arcs of G are mapped to half-lines meeting at $g(i) = 0$, and G is mapped onto a sector S of angle $3\pi/2$ (see 10.7 and 10.14). By conformality $g(0) = -1$ must lie on the bisector of the complementary sector, so S is as shown in Fig. 10.10. Note how we have avoided having to compute the angles subtended by the boundary arcs of G.

Stage 2 Working from the opposite end, we can realize $\hat{G} = \{w : |w| > 1\}$ as the image under the conformal map

$$h : \tau \mapsto \frac{\tau+1}{\tau-1} = w$$

of the right half-plane $\hat{S} = \{\tau : |\tau - 1| < |\tau + 1|\}$.

Stage 3 To transform S onto $\hat{S}$ we seek to multiply angles at 0 by a factor of $\frac{2}{3}$. Formally $\zeta \mapsto \tau = \zeta^{\frac{2}{3}}$ provides the map we want, but

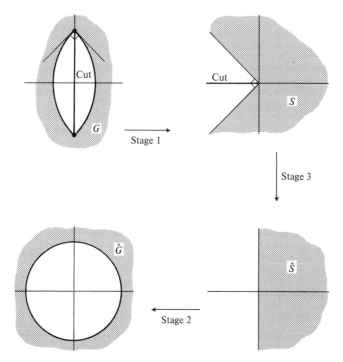

Fig. 10.10

we have to take care because this is a multifunction. We start with the z-plane cut along $[-i, i]$. In this cut plane there exists a holomorphic branch k of $[[(z-i)/(z+i)]^{\frac{2}{3}}]$. The map we finally require is $f = h \circ k$. This sends z to w where $[(z-i)/(z+i)]^2 = \tau^3$ and $(\tau+1)/(\tau-1) = w$. Hence $f(z) = w$, where

$$\left(\frac{z-i}{z+i}\right)^2 = \left(\frac{w+1}{w-1}\right)^3. \qquad \square$$

10.21 Remarks

Because of the way they act on circlines, Möbius transformations, powers, and exponentials are of most use for mapping regions whose boundaries are made up of circular arcs, lines, and line segments. However their scope is, even so, limited. For example, to handle regions with polygonal boundaries, one must introduce the *Schwarz-Christoffel transformation*, which is awkwardly defined by an integral:

$$z \mapsto \int_0^z (\zeta - z_1)^{-k_1} (\zeta - z_2)^{-k_2} \ldots (\zeta - z_n)^{-k_n} \, d\zeta.$$

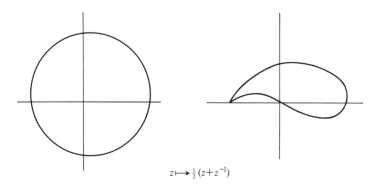

$$z \longmapsto \tfrac{1}{2}(z+z^{-1})$$

Fig. 10.11

For suitable z_j and k_j and the integrand a suitably specified holomorphic branch, this can be shown to map an open disc onto a region bounded by an n-gon.

Variants on the Joukowski transformation introduced in 10.17 are useful in elementary mapping problems of the type considered already, but also have a particular feature which makes them of interest in fluid dynamics: they map certain circles to (crude models of) aerofoil shapes (see Fig. 10.11). This enables the lift on a model of an aircraft wing to be estimated.

It is by no means obvious that it is possible to construct a conformal mapping from a region with a complicated, spiky boundary onto a civilized region such as $D(0; 1)$, or vice versa. The definitive theorem on conformal mapping, which we state without proof, is thus very striking.

The Riemann mapping theorem Let G be a simply connected region with $G \neq \mathbb{C}$. Then there exists a one-to-one conformal mapping f from G onto $D(0; 1)$ with $f^{-1}: D(0; 1) \to G$ also conformal.

It is worth noting that in each of our worked examples, the function we defined not only took one prescribed region, G, onto another, $\hat{G}$, but also mapped the boundary of G onto the boundary of $\hat{G}$. This extension of a conformal mapping to the boundary of a region is needed in many applications (see 10.39). In general, whether it is possible depends on the topological nature of the boundary.

Holomorphic mappings: some theory

It is often necessary to construct a conformal mapping $f: G \to \hat{G}$ such that the inverse mapping $f^{-1}: \hat{G} \to G$ exists and is also conformal; see 10.38. We present a group of theorems which have a bearing on this problem and are of independent interest. Since there are common themes in the proofs we begin with some general remarks.

10.22 Observations

Suppose G is open, $f \in H(G)$, and $a \in G$.

(1) Assume that G is a region and f non-constant. Then $f - f(a)$ is never zero in some $D'(a; r)$ (by the Identity theorem, 5.14).

(2) Let f be one-to-one. Then f' has isolated zeros (by Stage 1 of the Identity theorem proof, applied to f'; see 5.14).

(3) Choose r such that $\bar{D}(a; r) \subseteq G$ and suppose $f - f(a)$ is non-zero on γ^*, where $\gamma = \gamma(a; r)$. Let $m := \inf\{|f(z) - f(a)| : z \in \gamma^*\}$. Then

(i) $m > 0$ (by 1.18);

(ii) for each $w \in D(f(a); m)$, $f - f(a)$ and $f - w$ have the same number of zeros in $D(a; r)$ (counted according to multiplicity) (by Rouché's theorem, 7.7: for $z \in \gamma^*$, $|f(z) - f(a)| \geq m > |f(a) - w| = |f(a) - f(z) + f(z) - w|$).

10.23 Theorem

Suppose f is holomorphic and one-to-one in an open set G. Then f is conformal in G.

Proof. Assume for a contradiction that there exists $a \in G$ such that $f'(a) = 0$. Choose r such that $\bar{D}(a; r) \subseteq G$ and f' is never zero in $D'(a; r)$. This is possible by 10.22(2). Let $w \in D'(f(a); m)$, where m is as in 10.22(3). By 10.22(3)(ii), $f - f(a)$ and $f - w$ have the same number of zeros in $D(a; r)$. The function $f - f(a)$ has a zero of order at least two at a (by 6.8). On the other hand, $f - w$ cannot have two distinct zeros, since f is one-to-one, and cannot have a zero of order greater than one, since $f - w$ and $(f - w)'$ cannot both be zero at any point in $D(a; r)$. $\qquad\square$

10.24 The Open mapping theorem

Suppose f is holomorphic and non-constant in an open set G. Then $f(G)$ is open.

Proof. Fix $a \in G$. Choose r and m as in 10.22(3). By 10.22(1) and 10.22(3)(ii), $f - w$ has at least one zero in $D(a; r)$ whenever $w \in D(f(a); m)$. Hence $f(a) \in D(f(a); m) \subseteq f[D(a; r)] \subseteq f(G)$, and the result follows from the definition of an open set. □

10.25 The Inverse-function theorem

Let G be open and let f be holomorphic and one-to-one in G. Then f^{-1} is holomorphic in $f(G)$.

Proof. Let $b = f(a) \in f(G)$. Then $a = g(b)$, where $g := f^{-1}$. The Open mapping theorem implies that g is continuous (apply it to f in $D(a; \varepsilon) \subseteq G$ to obtain $\delta > 0$ such that $D(b; \delta) \subseteq f[D(a; \varepsilon)] = g^{-1}[D(a; \varepsilon)]$, which is the $\varepsilon - \delta$ definition of continuity of g at a, in shorthand.

By Theorem 10.23, $f'[g(b)] \neq 0$. Then

$$\frac{g(w) - g(b)}{w - b} = \frac{g(w) - g(b)}{f[g(w)] - f[g(b)]} \to \frac{1}{f'[g(b)]} \quad \text{as } g(w) \to g(b),$$

and hence also as $w \to b$, since we have proved that g is continuous. □

10.26 Remarks

(1) There is a partial converse to 10.23: if f is conformal in a region G, then f is locally one-to-one (Exercise 10.12).
(2) Suppose G is a region and $f : G \to f(G) := \hat{G}$ is one-to-one (and hence conformal by 10.23). By the Inverse-function theorem, $f^{-1} : \hat{G} \to G$ is also conformal.

The next result greatly improves on that obtained in 10.13(2).

10.27 Example

Suppose $f : D(0; 1) \to D(0; 1)$ is one-to-one, onto, and conformal, and maps $\alpha \in D(0; 1)$ to 0. Prove that, for some real constant λ,

$$f(z) = e^{i\lambda} \phi_\alpha(z) \ (z \in D(0; 1)), \quad \text{where } \phi_\alpha(z) := (z - \alpha)/(\bar{\alpha}z - 1).$$

Solution. Let $h = f \circ \phi_\alpha$. It can easily be shown that h maps $D(0; 1)$ one-to-one onto $D(0; 1)$. Since $h(w)/w$ has a removable singularity at 0, there exists $g \in H(D(0; 1))$ such that $wg(w) = h(w)$ for all $w \in D(0; 1)$. On $|w| = r < 1$, $|g(w)| \leq |h(w)|/r < 1/r$. Applying the Maximum-modulus theorem, 5.20, to g, we have $|g(w)| \leq 1/r$

whenever $|w| \leq r < 1$ and hence

$$|h(w)| \leq |w| \quad \text{for all } w \in D(0; 1).$$

The same argument applies to h^{-1} (thanks to the Inverse-function theorem) and gives

$$|w| \leq |h(w)| \quad \text{for all } w \in D(0; 1).$$

Combining these inequalities we see that $|g(w)| = 1$ for all $w \in D(0; 1)$. But then 2.6(2) implies that g is a constant of modulus one. Hence for some real constant λ, $f[\phi_\alpha(w)] = e^{i\lambda}w$ for all $w \in D(0; 1)$. The result follows once we note that $\phi_\alpha = \phi_\alpha^{-1}$. □

Harmonic functions

Our treatment of harmonic functions is self-contained, but we do not give proofs in great detail, anticipating that this section will be of most interest to those who have already met harmonic functions in some other setting. We concentrate on the relationship between holomorphic and harmonic functions, and on the use of conformal mapping.

10.28   Definition

Let G be an open subset of $\mathbb{C}$, and identify $z = x + iy \in \mathbb{C}$ with $(x, y) \in \mathbb{R}^2$. A function $u : G \to \mathbb{R}$ is *harmonic in G* if
   (i) u has continuous second order partial derivatives in G, and
   (ii) u satisfies Laplace's equation $u_{xx} + u_{yy} = 0$ in G.
We write $\mathcal{H}(G)$ to denote the set of functions harmonic in G.

10.29   Theorem

Let f be holomorphic in an open set G, where

$$f(z) = u(x, y) + iv(x, y) \quad \text{(for } z = x + iy \in G\text{)}.$$

Then u and v are harmonic in G.

Proof. We know from the proof of Theorem 2.4 that

$$f'(z) = u_x + iv_x = -iu_y + v_y,$$

where u_x, u_y, v_x, and v_y denote the partial derivatives of u and v at $z = x + iy$. Hence come the Cauchy–Riemann equations $u_x = v_y$, $u_y = -v_x$. We can also deduce that u and v have partial derivatives of all orders (because f is infinitely differentiable) and so certainly have continuous second order partial derivatives. This ensures that $u_{xy} = u_{yx}$ and $v_{xy} = v_{yx}$. Then

$$u_{xx} = v_{yx} = v_{xy} = -u_{yy} \quad \text{and} \quad v_{xx} = -u_{yx} = -u_{xy} = -v_{yy}. \qquad \square$$

To reveal fully the relationship between complex analysis and harmonic functions we need a (partial) converse to Theorem 10.29. We first prove a technical lemma.

10.30 Lemma

Suppose that $f(z) = u(x, y) + iv(x, y)$ for $z = x + iy$ in an open set G, where u and v have continuous first order partial derivatives and satisfy the Cauchy–Riemann equations in G. Then $f \in H(G)$.

Proof. Let $z \in G$, $D(z; r) \subseteq G$, and take $h = p + iq$ such that $|h| < r$. Then

$$\frac{f(z+h) - f(z)}{h} = \frac{p}{h} \left(\frac{u(x+p, y+q) - u(x, y+q)}{p} \right.$$
$$\left. + i \frac{v(x+p, y+q) - v(x, y+q)}{p} \right)$$
$$+ \frac{q}{h} \left(\frac{u(x, y+q) - u(x, y)}{q} + i \frac{v(x, y+q) - v(x, q)}{q} \right)$$
$$= \frac{p}{h} \left(\frac{\partial u}{\partial x}(x + \alpha p, y+q) + i \frac{\partial v}{\partial x}(x + \beta p, y+q) \right)$$
$$+ \frac{q}{h} \left(\frac{\partial u}{\partial y}(x, y + \gamma q) + i \frac{\partial v}{\partial y}(x, y + \delta q) \right),$$

$$\text{where} \quad 0 < \alpha, \beta, \gamma, \delta < 1,$$

by the Mean-value theorem. Using the continuity of the partial derivatives and the given Cauchy–Riemann equations, we see that $f'(z)$ exists. $\qquad \square$

10.31 Theorem

Let G be an open disc [a simply connected region], and suppose $u \in \mathcal{H}(G)$. Then there exists $f \in H(G)$ such that $\operatorname{Re} f = u$. (The function f is sometimes called a *complex potential*, and $\operatorname{Im} f$ a *harmonic conjugate* for u.)

Proof. If f exists we must have $f'(z) = u_x - iu_y =: g(z)$. We therefore seek an antiderivative for g. We apply Lemma 10.30 to g to show $g \in H(G)$. The Antiderivative theorem, 4.4 [4.14], provides $F \in H(G)$ with $F' = g$. Then $(u - \mathrm{Re}\, F)_x = (u - \mathrm{Re}\, F)_y = 0$ in G, whence $u - \mathrm{Re}\, F$ is a real constant, k (cf. 3.18). Now let $f = F + k; f \in H(G)$ and $\mathrm{Re}\, f = u$. $\qquad \square$

Remark In practice, given G and u, f can often be recognized at sight, from $f'(z) = u_x - iu_y$. (Suppose for example $u(x, y) = x - xy$, which is harmonic in $G = \mathbb{C}$. Then $u_x - iu_y = 1 - y + ix = 1 + iz$. Hence $u = \mathrm{Re}\, f$, where $f(z) = z + \frac{1}{2}iz^2$.) Where we cannot spot a solution for f, we use the relation

$$f(w) - f(a) = \int_{\gamma(w)} f'(z)\, dz = \int_{\gamma(w)} (u_x - iu_y)\, dz,$$

which holds for any polygonal path $\gamma(w)$ in G joining a to w; it is often convenient to take a path consisting of horizontal and vertical line segments.

We now use 10.31 to deduce theorems about harmonic functions from corresponding theorems about holomorphic functions.

10.32 The Poisson integral formula

Suppose $u \in \mathcal{H}(G)$, where G is an open disc containing $\bar{D}(0; 1)$. For $z = re^{i\theta} \in D(0; 1)$,

$$u(re^{i\theta}) = \frac{1}{2\pi} \int_0^{2\pi} \frac{(1 - r^2)}{(1 - 2r\cos(\theta - t) + r^2)}\, u(e^{it})\, dt.$$

Proof. Choose $f \in H(G)$ such that $u = \mathrm{Re}\, f$, by Theorem 10.31. The required formula is obtained from the Poisson integral formula for f, 5.8, by equating real parts. $\qquad \square$

10.33 The Mean-value property

Under the same hypotheses as in 10.32,

$$u(0) = \frac{1}{2\pi} \int_0^{2\pi} u(e^{it})\, dt.$$

Proof. The formula is clearly a special case of that in 10.32. It can be derived more directly from Cauchy's integral formula (of which it is the harmonic analogue) and Theorem 10.31. $\qquad \square$

Note The hypotheses in 10.32 and 10.33 can be weakened. It is enough to assume that u is harmonic in $D(0; 1)$ and continuous on $\bar{D}(0; 1)$. To see this, apply the preceding theorems to u_ρ, where $u_\rho(z) := u(\rho z)$ $(\rho < 1)$ and take the limit as ρ increases to 1.

10.34 The Maximum principle for harmonic functions

Let G be a bounded region, and let u be harmonic in G and continuous on $\bar{G}$. Suppose $u \leqslant M$ on $\partial G = \bar{G} \backslash G$, where M is a constant. Then $u \leqslant M$ on $\bar{G}$, that is, u attains its maximum on the boundary ∂G of G.

Sketch proof. The Maximum principle is the harmonic counterpart of the Maximum-modulus theorem, 5.20, and can be proved, *mutatis mutandis*, in the same way. In outline, a local version is first derived from the Mean-value property (cf. 5.19), and then $\{z \in G : u(z) = M\}$ is shown to be the whole of G or the empty set (cf. the proof of the Identity theorem, 5.14, Stage 2). □

10.35 The Dirichlet problem

Let G be a region. Suppose we are given a real-valued continuous function U on the boundary $\partial G = \bar{G} \backslash G$. Can we find a function u such that u is continuous on $\bar{G}$ and harmonic in G and such that $u = U$ on ∂G? This boundary value problem is known as the *Dirichlet problem*. (The solution is unique if it exists: if u_1 and u_2 are both solutions, apply the Maximum principle to $u_1 - u_2$ and $u_2 - u_1$ to prove $u_1 \equiv u_2$ on $\bar{G}$.) The simpler the geometric configuration, the simpler, presumably, is the problem. We now solve the Dirichlet problem for the simplest case of all: $G = D(0; 1)$.

10.36 The Dirichlet problem for a disc

Suppose U is a real-valued continuous function on the unit circle. Let

$$v(re^{i\theta}) := \frac{1}{2\pi} \int_0^{2\pi} \frac{(1 - r^2)}{(1 - 2r \cos(\theta - t) + r^2)} U(e^{it}) \, dt \qquad (re^{i\theta} \in D(0; 1))$$

be the *Poisson integral* of U. Define u by

$$u(re^{i\theta}) = \begin{cases} v(re^{i\theta}) & (0 \leqslant r < 1), \\ U(re^{i\theta}) & (r = 1). \end{cases}$$

Then (1) $u \in \mathcal{H}(D(0; 1))$, and (2) u is continuous on $\bar{D}(0; 1)$.

Sketch proof. The *Poisson kernel*

$$P_r(\theta - t) := \frac{1 - r^2}{1 - 2r \cos(\theta - t) + r^2}$$

may be alternatively written as $\mathrm{Re}[(w + z)/(w - z)]$ or as $(1 - |z|^2)/|w - z|^2$ (where $w = e^{it}$ and $z = re^{i\theta}$); see Exercise 1.6. For $|z| < 1$,

$$u(z) = \mathrm{Re}\!\left(\frac{1}{2\pi i} \int_{\gamma(0;1)} \frac{w + z}{w - z}\, U(w)\, \frac{dw}{w} \right)$$

$$= \mathrm{Re}\!\left(\frac{1}{\pi i} \int_{\gamma(0;1)} \frac{U(w)}{w - z}\, dw \right) - \mathrm{Re}\!\left(\frac{1}{2\pi i} \int_{\gamma(0;1)} \frac{U(w)}{w}\, dw \right).$$

The second term is constant, while the first is the real part of a holomorphic function; see Exercise 5.8. Hence u is harmonic, by 10.29.

To prove (2), we have to show that if $\alpha \in [0, 2\pi]$, then $v(re^{i\theta}) \to U(e^{i\alpha})$ as $z = re^{i\theta} \to e^{i\alpha}$. Putting u equal to the constant function 1 in the Poisson integral formula gives

$$1 = \frac{1}{2\pi} \int_0^{2\pi} P_r(\theta - t)\, dt.$$

Hence

$$v(re^{i\theta}) - U(e^{i\alpha}) = \frac{1}{2\pi} \int_0^{2} P_r(\theta - t)[U(e^{it}) - U(e^{i\alpha})]\, dt.$$

Let $\varepsilon > 0$. Continuity of U implies that there exists $\delta > 0$ such that $|U(e^{it}) - U(e^{i\alpha})| < \varepsilon$ for all t such that e^{it} lies on the arc J of the unit circle T containing $e^{i\alpha}$ and joining $e^{i(\alpha \pm \delta)}$. Let K be the arc of T complementary to J. Then for some $M > 0$, $|w - z| \geqslant M$ whenever $w \in K$ and z lies in the shaded sector (see Fig. 10.12).

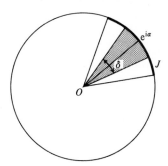

Fig. 10.12

We have, noting that $P_r(\theta - t) \geq 0$,

$$|v(re^{i\theta}) - U(e^{i\alpha})| \leq \frac{1}{2\pi} \int_0^{2\pi} P_r(\theta - t) |U(e^{it}) - U(e^{i\alpha})| \, dt.$$

Elementary estimates show that the contribution from J to this integral is less than ε, while, for z in the shaded sector, the contribution from K is bounded by a multiple of $(1 - r^2)/M^2$ (remember that U is bounded on the circle). Hence we can make $|v(re^{i\theta}) - U(e^{i\alpha})| < 2\varepsilon$ by taking $|re^{i\theta} - e^{i\alpha}|$ small enough. $\square$

The preceding results about $D(0; 1)$ adapt to an arbitrary disc $D(a; R)$ by translating and rescaling. Transfer to other regions may be accomplished using conformal mappings. The key to this lies in the next lemma.

10.37 Lemma

Suppose that G and $\hat{G}$ are open sets, that $g : G \to \hat{G}$ is holomorphic, and that $\hat{u} \in \mathcal{H}(\hat{G})$. Then $u := \hat{u} \circ g \in \mathcal{H}(G)$.

Proof. Put $\xi + i\eta = g(z) = g(x + iy)$, so that $\hat{u}(\xi, \eta) = u(x, y)$. Then straightforward partial differentiation together with the formulae for the derivative in 2.4 show that

$$u_{xx} + u_{yy} = |g'(z)|^2 (\hat{u}_{\xi\xi} + \hat{u}_{\eta\eta}) = 0. \qquad \square$$

10.38 The solution of the Dirichlet problem by conformal mapping

To illustrate the use of the lemma, suppose we have a one-to-one continuous mapping g of $\bar{G}$ onto $\bar{D}(0; 1)$ which maps G conformally onto $D(0; 1)$; $g^{-1} : D(0; 1) \to G$ is holomorphic by the

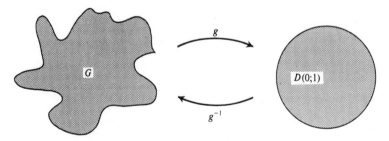

Fig. 10.13

Inverse-function theorem, 10.25. Let U be a real-valued continuous function on ∂G. Let $\hat{u}$ be the solution of the Dirichlet problem for $D(0; 1)$ with boundary values given by $U \circ g^{-1}$. Then $\hat{u} \circ g$ is continuous on $\bar{G}$, harmonic in G, and equals U on ∂G. In practical problems the boundary function U is frequently piecewise continuous, rather than continuous. Theorem 10.36 can be extended to cover this case.

10.39 Examples

(1) Let $G = \{z : 0 < \operatorname{Re} z < \pi, \operatorname{Im} z > 0\}$. To find: a function u such that

   (i) u is continuous on $\bar{G}$ except at 1, and harmonic in G;
   (ii) $u(z) = -1$ when $\operatorname{Re} z = \pi$ $(\operatorname{Im} z \geqslant 0)$ and when $\operatorname{Im} z = 0$ $(0 < \operatorname{Re} z \leqslant \pi)$;
   (iii) $u(z) = 0$ when $\operatorname{Re} z = 0$ $(\operatorname{Im} z > 0)$.

Solution. We can use the map $g : z \mapsto w = \cos z$ to map $\bar{G}$ one-to-one onto the closed lower half-plane; G is then mapped conformally onto the open lower half-plane (see Example 10.19). The boundary of G is mapped to the real axis, with $g(0) = 1$, $g(\pi) = -1$.

Define $u(w) = \dfrac{1}{\pi} \arg(w - 1)$, where arg takes values in $[-\pi, \pi]$. As the imaginary part of a holomorphic branch of the logarithm, this is harmonic, and it takes the value zero on $(1, \infty)$ and -1 on $(-\infty, 1)$. A suitable choice for u is then

$$u(z) = \frac{1}{\pi} \arg(\cos z - 1) \quad (z \in \bar{G}). \qquad \square$$

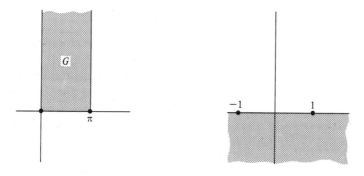

Fig. 10.14

(2) The Dirichlet problem for a half-plane Let G be the open upper half-plane and suppose U is continuous on the real axis. Fix $z = x + iy$ $(y > 0)$. For Im $\zeta > 0$, define $g(\zeta) = \dfrac{\zeta - z}{\zeta - \bar{z}}$, so that g is as in 10.38. The solution, u, of the Dirichlet problem for the upper half-plane satisfies $u(z) = \hat{u}[g(z)] = \hat{u}(0)$, where $\hat{u}$ is given by the Poisson integral of $U \circ g^{-1}$. We have

$$e^{it} = g(\tau) = \frac{\tau - z}{\tau - \bar{z}} \quad (t \in [0, 2\pi], \tau \in \mathbb{R}),$$

so that formally

$$ie^{it}\, dt = \frac{z - \bar{z}}{(\tau - \bar{z})^2}\, d\tau,$$

whence

$$dt = \frac{2y}{|\tau - \bar{z}|^2}\, d\tau.$$

We find that

$$u(x, y) = u(z) = \frac{y}{\pi} \int_{-\infty}^{\infty} \frac{U(\tau)}{(\tau - x)^2 + y^2}\, d\tau.$$

This is, of course just the answer we got in 9.24 for this problem. However, the method here depends on the change of variables from t to τ, which we have not justified.

Postscript

We have studied harmonic functions as, locally, the real parts of holomorphic functions. We have made little direct use of their definition as suitably smooth solutions of Laplace's equation in two dimensions. Many readers will have seen this equation before, either in the context of partial differential equations or in connection with a mathematical model for 2-dimensional problems in fluid flow, heat flow, or electrostatics. Such results as the Mean-value property and the Maximum principle may well be familiar. These are usually derived using the apparatus of vector calculus; the proofs have their roots in Green's theorem (or, more fundamentally, in Stokes' theorem on differential forms). It should perhaps now come as no surprise that Cauchy's theorem and Green's theorem are intimately related. Take a function $f = u + iv$

holomorphic inside and on a contour γ. Blithely ignoring any possible technical problems, we may realize $\int_\gamma f(z)\,dz$ as the sum of line integrals

$$\int_\gamma (u\,dx - v\,dy) + i\int_\gamma (u\,dy + v\,dx)$$

(by putting $dz = dx + i\,dy$). We may then use Green's theorem to rewrite this as

$$\iint_{I(\gamma)} (-v_x - u_y)\,dx\,dy + i\iint_{I(\gamma)} (u_x - v_y)\,dx\,dy,$$

which is zero by the Cauchy–Riemann equations. So $\int_\gamma f(z)\,dz = 0$. This is not the short cut it might seem. To justify it one needs to assume (or somehow prove otherwise) that f' is continuous, a fact we deduced from a version of Cauchy's theorem. The approach is certainly of historical interest: it was the one used by Cauchy to derive the theorem that bears his name.

Exercises

1. Find the image of (a) $\{z : 0 < \arg z < \pi/6\}$, (b) $D(0; 2)$, (c) $\{z : 0 < \operatorname{Im} z < 1\}$ under each of the maps (i) $z \mapsto (1 + i)z$, (ii) $z \mapsto z^2$, (iii) $z \mapsto 1/z$.

2. Let $f(z) = 2iz/(z + i)$. Prove that f maps circular arcs through 0 and i to themselves and deduce that f maps $\{z : \operatorname{Re} z > 0, |z - \tfrac{1}{2}i| < \tfrac{1}{2}\}$ to itself. What is the image under f of $\{z : |z| < |z - i|\}$?

3. (i) Determine those circles $|z - a| = r$ which have -3 and 1 as inverse points.

(ii) Determine the circline which has 0 and $1 + i$ as inverse points and which passes through 1.

4. Describe the image of
   (i) $\{z : |z - 1| > 1\}$ under $z \mapsto w = z/(z - 2)$,
   (ii) $\{z : 0 < \arg z < \pi/4\}$ under $z \mapsto w = z/(z - 1)$,
   (iii) $\{z : 0 < \arg z < \pi/2\}$ under $z \mapsto w = (z - 1)/(z + 1)$,
   (iv) $\{z : \operatorname{Re} z > 0\}$ under $z \mapsto w$, where $(w - 1)/(w + 1) = 2(z - 1)/(z + 1)$,
   (v) $\{z : |z - 1| > 1$ and $|z + 1| > 1\}$ under $z \mapsto w = i(z + 2)/z$,
   (vi) $\{z : 1/2 < |z| < 1\}$ under $z \mapsto w = (2z + 1)/(z - 2)$.

5. Find Möbius transformations to map
   (i) $1, i, 0$ to $1, i, -1$, respectively,
   (ii) $0, 1, \infty$ to $\infty, -i, 1$, respectively,
   (iii) $-1, \infty, i$ to $0, \infty, 1$, respectively.
In each case describe the circlines whose images are straight lines.

6. Find the Möbius transformation mapping 0, 1, and ∞ to 1, $1+i$, and i, respectively. Under this mapping what is the image of
  (i) a circular arc through -1 and $-i$,
  (ii) the line given by Im $z =$ Re z,
  (iii) the real axis,
  (iv) the imaginary axis?

7. Describe the Möbius transformations mapping the open upper half-plane onto $D(0; 1)$ which map the imaginary axis onto the real axis.

8. Find a common pair of inverse points for the circles $|z| = 1$ and $|z - 1| = \frac{5}{2}$. Hence find a conformal mapping of the region bounded by these two circles onto an annulus.

9. Find the image of
  (i) $\{z : 0 < \arg z < \pi/4\}$ under $z \mapsto w = iz^4$,
  (ii) $\{z : 0 < \text{Re } z < 1, 0 < \text{Im } z < \pi/2\}$ under $z \mapsto w = e^z$,
  (iii) $\{z : |z| < 1, 0 < \arg z < \pi/3\}$ under $z \mapsto w = \left(\dfrac{z^3 + 1}{z^3 - 1}\right)^2$,
  (iv) $\{z : \text{Re } z > -1\}$ under $z \mapsto w = z^2$,
  (v) $\{z : 0 < \text{Im } z < \pi\}$ under $z \mapsto w = (1 + ie^z)/(1 - ie^z)$.

10. Construct a conformal mapping onto $D(0; 1)$ of each of the following:
  (i) $\{z : -\frac{1}{4}\pi < \arg z < \frac{1}{4}\pi\}$, (iii) $\{z : \text{Re } z > 0, \text{Im } z > 0, |z| > 1\}$,
  (ii) $\{z : -1 < \text{Re } z < 1\}$, (iv) $\{z : \text{Re } z > 0 \text{ or } \text{Im } z \neq 0\}$.

11. Given that $-1 < c < 1$, find a conformal mapping of $\{z : |z| < 1,$ Re $z > c\}$ onto the open upper half-plane.

12. Use the facts given in 10.22 to prove that if f is conformal in a region G, then, for each $a \in G$, there exists $r > 0$ such that the restriction of f to $D(a; r)$ is one-to-one.

13. Check that each of the following functions is harmonic on the indicated set, and find holomorphic function of which it is the real part:
(i) $x^2 - y^2 + x$ (on $\mathbb{C}$) (ii) $x - y(x^2 + y^2)^{-1}$ (on $\mathbb{C}\backslash[0, \infty)$), (iii) $\sin(x^2 - y^2)e^{-2xy}$ (on $\mathbb{C}$), (iv) $\log(x^2 + y^2)^{\frac{3}{2}}$ (on the open first quadrant).

14. Find a conformal mapping of $G = \{z : 0 < \arg z < 3\pi/2\}$ onto a strip. Hence find a function u which is continuous on $\bar{G}$ except at 0, which is harmonic in G, and which is such that $u(x, 0) = 1$ $(x > 0)$ and $u(0, y) = 0$ $(y < 0)$.

15. Suppose u is harmonic in $\mathbb{C}$ and $u \geq 0$. Show that, for $r < R$.

$$u(re^{i\theta}) = \frac{1}{2\pi} \int_0^{2\pi} \frac{R^2 - r^2}{R^2 - 2rR\,\cos(\theta - t) + r^2} u(Re^{it})\,dt.$$

Deduce that, if $r < R$,

$$\frac{R - r}{R + r} u(0) \leq u(re^{i\theta}) \leq \frac{R + r}{R - r} u(0).$$

Hence show that a bounded function which is harmonic in $\mathbb{C}$ is necessarily constant.

Supplementary exercises

Chapter 1

1. Compute $\mathrm{Im}[(1 - 2i)(1 + 2i)]$, $\overline{(2 + i)^{-1}}$, and $|(1 - 2i)(2 - i)|$.

2. Let $z, w \in \mathbb{C}$. Prove that

$$|z + iw|^2 + |w + iz|^2 = 2(|z|^2 + |w|^2)$$

and deduce that

$$|z + iw|^2 \leqslant 2(|z|^2 + |w|^2).$$

3. Evaluate $\sum_{k=0}^{n} e^{ik\theta}$. Deduce that

$$1 + 2 \sum_{k=1}^{n} \cos k\theta = \frac{\sin(n + \frac{1}{2})\theta}{\sin \frac{1}{2}\theta} \quad \text{if } \theta \neq 2m\pi \quad (m \in \mathbb{Z}).$$

What is the value of $\sum_{k=1}^{n} \sin k\theta$?

4. Describe the following loci:

   (i) $|z + i| = |z - 3i|$,   (iv) $|\mathrm{Re}(z + 1)| = |z - 1|$,
   (ii) $|z + 1| = 4|z - 1|$,   (v) $\mathrm{Re}\, z^2 = 1$.
   (iii) $|z - 1| + |z + 1| = 3$,

5. Let S be a finite subset of $\mathbb{C}$.
   (i) Prove that S is open only if $S = \varnothing$.
   (ii) Prove that S is closed.

6. Let a be any complex number. Let $z_0 = a$ and, for $n \geqslant 1$, define

$$z_{n+1} = \frac{1}{2}\left(z_n - \frac{1}{z_n}\right),$$

if $z_n \neq 0$. Prove the following assertions.
   (i) If $\langle z_n \rangle$ converges to a limit z, then $z^2 + 1 = 0$.
   (ii) If a is real, then $\langle z_n \rangle$, if defined, does not converge.
   (iii) If $a = ib$, where $b \in \mathbb{R} \backslash \{0\}$, then $\langle z_n \rangle$ converges.
   (iv) If $|a| = 1$ and $a \neq \pm 1$, then $\langle z_n \rangle$ converges.
   (v) If $\mathrm{Im}\, a > 0$, then $\langle z_n \rangle$ converges to i and if $\mathrm{Im}\, a < 0$, then $\langle z_n \rangle$ converges to $-i$. (Hint: consider $|z_{n+1} - i|/|z_{n+1} + i|$ and use 1.6(1).)

7. Suppose that S is a compact subset of $\mathbb{C}$ and $f : S \to \mathbb{C}$ a continuous function. Use 1.18 to prove that the image, $f(S)$, of S under f is compact.

8. Let $S = \{z \in \mathbb{C} : |z| = 1\}$ and let $f : S \to \mathbb{R}$ be continuous. Prove that there exists $w \in S$ such that $f(w) = f(-w)$.

Chapter 2

1. Let f be holomorphic in $D(0; 1)$.
(i) Define g by $g(z) = \overline{f(\bar{z})}$. Prove that g is holomorphic in $D(0; 1)$.
(ii) Define k by $k(z) = \overline{f(z)}$. Prove that k is differentiable at a if and only if $f'(a) = 0$. Deduce that k is holomorphic in $D(0; 1)$ only if f is constant.

2. Prove that $z/(1 + |z|)$ is not holomorphic anywhere. (cf. Exercise 1.16.)

3. Let $f(z) = z^3$. Prove that there exists no point c on the line segment $[1, i]$ such that
$$\frac{f(i) - f(1)}{i - 1} = f'(c).$$

(Thus the Mean-value theorem does not extend to holomorphic functions.)

4. Determine for which values of $z \in \mathbb{C}$ the following series converge:

$$\text{(i)} \ \sum (n + z)^{-1} \qquad \text{(ii)} \ \sum (n + z)^{-2}.$$

5. Obtain power series expansions for $(1 + z)^{-2}$ and for $(1 + z)^{-3}$, each valid for $|z| < 1$. (Hint: use 2.12.)

Let $p(z)$ be a polynomial of degree $k > 0$. Prove that $\sum p(n)z^n$ has radius of convergence 1 and that there exists a polynomial $q(z)$ of degree k such that

$$\sum_{n=0}^{\infty} p(n)z^n = q(z)(1 - z)^{-(k+1)} \qquad (|z| < 1).$$

6. Suppose $\omega^3 = 1$, $\omega \neq 1$. Express $e^z + e^{\omega z} + e^{\omega^2 z}$ as a power series. Hence evaluate $\sum_{n=0}^{\infty} 8^n/(3n)!$. Find also $\sum_{n=0}^{\infty} 27^n/(3n + 1)!$.

7. Let $p(z) = (z - \alpha_1)(z - \alpha_2) \ldots (z - \alpha_k)$, where $\alpha_1, \alpha_2, \ldots, \alpha_k$ are distinct complex numbers. Let $M = \min_{1 \leq r \leq k} |\alpha_r|$. Prove that, for $|z| < M$, it is possible to represent $p(z)^{-1}$ as a power series. Could the radius of convergence of this series exceed M?

8. Prove that, for $z = x + iy$,

$$|\cos z|^2 = \tfrac{1}{2}(\cosh 2y + \cos 2x) = \sinh^2 y + \cos^2 x = \cosh^2 y - \sin^2 x,$$
$$|\sin z|^2 = \tfrac{1}{2}(\cosh 2y - \cos 2x) = \sinh^2 y + \sin^2 x = \cosh^2 y - \cos^2 x.$$

Deduce that $|\cos z|^2 + |\sin z|^2 = 1$ if and only if z is real.

9. Prove that

$$\tan(x + iy) = \frac{\sin 2x + i \sinh 2y}{\cos 2x + \cosh 2y}.$$

10. Show that if $w \in [z^{\alpha+\beta}]$, then there exist $z_1 \in [z^{\alpha}]$ and $z_2 \in [z^{\beta}]$ such that $w = z_1 z_2$.

Give an example to show that it may happen that $z_1 \in [z^{\alpha}]$ and $z_2 \in [z^{\beta}]$ yet $z_1 z_2 \notin [z^{\alpha+\beta}]$. (Hint: try $\alpha + \beta = 0$.)

Chapter 3

1. Let $p(z)$ be a polynomial. Use the Fundamental theorem of calculus to prove that $\int_\gamma p(z)\,dz = 0$ for every closed path γ in $\mathbb{C}$. (This is a very special case of Cauchy's theorem (the subject of Chapter 4).)

Deduce that there exists $\varepsilon > 0$ such that, for every polynomial $p(z)$,

$$\left| p(z) - \frac{1}{z} \right| \geqslant \varepsilon \text{ whenever } |z| = 1.$$

[Thus the function $1/z$ cannot be uniformly approximated on the unit circle by polynomials.]

2. Let γ be a square contour such that γ^* has its vertices at $(\pm 1 \pm i)R$. Obtain an upper bound for $|\int_\gamma z^n\,dz|$ when (i) $n \in \mathbb{Z}$, $n \geqslant 0$, (ii) $n \in \mathbb{Z}$, $n < 0$.

3. Given a path γ, define length(γ) as in 3.10. Prove the following assertions.

(i) length($-\gamma$) = length(γ).
(ii) if γ is the join of paths γ_1 and γ_2, then

$$\text{length}(\gamma) = \text{length}(\gamma_1) + \text{length}(\gamma_2).$$

(iii) If $\bar{\gamma}$ is obtained by reparametrizing γ as in 3.5(3), then length($\bar{\gamma}$) = length(γ).

4. Given a closed path γ, define the *signed area* enclosed by γ to be

$$S = \frac{1}{2i} \int_\gamma \bar{z} \, dz.$$

By writing $\gamma(t)$ in terms of its real and imaginary parts, prove that S is real. Show that $|S|$ agrees with the usual area when γ is a circular contour or a triangular contour, and comment on the sign of S.

5. Suppose that $f(z) = \sum_0^\infty c_n z^n$ in $D(0; R)$. Prove that

$$\frac{1}{2\pi} \int_0^{2\pi} |f(re^{i\theta})|^2 \, d\theta = \sum_{n=0}^\infty |c_n|^2 r^{2n} \qquad (0 \le r < R).$$

(Hint: use 1.3(3) and then 3.13 twice, with the aid of 1.18(1).)

6. Are the following statements true or false? In each case give a proof or counterexample as appropriate.
    (i) $\{z : 1 < |z| < 2\}$ is a region.
    (ii) $\{z : |z - 1| < 1 \text{ or } |z + 1| < 1\}$ is a region.
    (iii) If G is a region and $D(a; r)$ is a proper subset of G, then $G \backslash D(a; r)$ is a region.
    (iv) If $G_1, \ldots, G_N$ are regions and $G_k \cap G_{k+1} \ne \varnothing$ for $k = 1, \ldots, N-1$ then $\bigcup_{k=1}^N G_k$ is a region.

7. Prove the assertion in 3.21(3) that $\mathbb{C}_\alpha$ is simply connected. (A possible strategy: given a closed path γ with $\gamma^* \subseteq \mathbb{C}_\alpha$ (i) find a point a such that $[b, z] \subseteq \mathbb{C}_\alpha$ for all $z \in \gamma^*$ and for all $b \in D(a; r)$, for some r; (ii) show, with the aid of 3.23, that there exist finitely many open convex sets $G_0, \ldots, G_{N-1}$ such that $a \in \bigcap_{k=1}^{N-1} G_k$ and γ is the join of paths γ_k for $0 \le k \le N-1$ where $\gamma_k^* \subseteq G_k$ (hint: each G_k may be taken to be an open sector bounded by a circular arc and two line segments meeting at a point in $D(a; r)$); (iii) deduce that γ is homotopic to the null path with image $\{a\}$.)

Chapter 4

1. Which of the following integrals are zero: (i) $\int_{\gamma(0;2)} z/|z| \, dz$, (ii) $\int_{\gamma(0;1)} \bar{z} \, dz$, (iii) $\int_\gamma z^5 \sin z \, dz$, where γ is as in Exercise 3.1(iv), (iv) $\int_{\gamma(2\pi;\pi)} \sec^2 z \, dz$?

2. A subset S of $\mathbb{C}$ is said to be *star-shaped* if there exists $a \in S$ such that $[a, z] \subseteq S$ for all $z \in S$.
    (i) Prove that a convex set is star-shaped, and exhibit a non-convex set which is star-shaped.
    (ii) Prove that an open star-shaped set is a region.

(iii) Let G be open and star-shaped and let $f \in H(G)$. Adapt the proof of 4.3 to prove that $\int_\gamma f(z) \, dz = 0$ (that is, prove Cauchy's theorem for a star-shaped region).

3. Which of the following sets are star-shaped: (i) $\{z : 1 < |z| < 2\}$, (ii) $\{z : |z - 2| > 3, \ |z| < 2\}$, (iii) $\mathbb{C}_\alpha$ (as defined in 3.21(3)), (iv) $\{z : \text{Im } z > 0, \ 1 < |z| < 2\}$, (v) $\mathbb{C}\setminus\{\pm 1\}$, (vi) $\mathbb{C}\setminus\{z : |z| = 1, \ \text{Re } z \geq 0\}$? Give reasons.

The remaining exercises concern Cauchy's theorem, Level II.

4. Describe γ^* for each of the following closed paths γ:
 (i) γ is the join of γ_1, γ_2, and γ_3, where $\gamma_k(t) = (\frac{5}{2} - k) + k e^{it}$ $(t \in [0, 2\pi])$,
 (ii) γ is the join of $[-2, -1]$, $-\Gamma_1$, $[1, 2]$, and Γ_2, where $\Gamma_r(t) = r e^{it}$ $(0 \leq t \leq \pi)$,
 (iii) γ is the join of $[-5, -1]$ and γ_1, where $\gamma_1(t) = \dfrac{t}{\pi} e^{it}$ $(t \in [\pi, 5\pi])$.

    Use 4.16 to compute $n(\gamma, 0)$ in each case.

5. Let $\gamma_1, \gamma_2 : [\alpha, \beta] \to \mathbb{C}$ be closed paths and define $\gamma(t) = \gamma_1(t)\gamma_2(t)$ and $\Gamma(t) = \gamma_1(t) + \gamma_2(t)$ $(t \in [\alpha, \beta])$.
 (i) Show that γ and Γ are closed paths.
 (ii) Show that if $0 \notin \gamma_1^* \cup \gamma_2^*$,

$$n(\gamma, 0) = n(\gamma_1, 0) + n(\gamma_2, 0).$$

(iii) Show that if $|\gamma_1(t)| > |\gamma_2(t)|$ for $t \in [\alpha, \beta]$, then $n(\Gamma, 0) = n(\gamma_1, 0)$.

Chapter 5

1. Let $a, b \in \mathbb{C}$, with $|a| \neq 1, |b| \neq 1$. Evaluate, distinguishing cases,

 (i) $\displaystyle\int_{\gamma(0;1)} \left(\frac{z - b}{z - a}\right)^2 dz,$ (ii) $\displaystyle\int_{\gamma(0;1)} \frac{1}{(z - a)(z - b)} \, dz.$

2. Evaluate $\int_{\gamma(0;1)} \dfrac{\cos z}{z} \, dz$. Deduce that

$$\int_0^{2\pi} \cos(\cos \theta)\cosh(\sin \theta) \, d\theta = 2\pi.$$

3. Suppose that f is holomorphic inside and on $\gamma(0; 1)$. Let $u = \text{Re } f$ and $v = \text{Im } f$.
 (i) Evaluate the integrals

$$\frac{1}{2\pi} \int_0^{2\pi} (u(e^{it}) \pm v(e^{it})) e^{-int} \, dt \quad (n \geq 0),$$

in terms of the coefficients of the Taylor series for f.

(ii) Deduce that, if $f(0)$ is real,

$$f(z) = \frac{1}{2\pi} \int_0^{2\pi} \left(\frac{e^{it} + z}{e^{it} - z} \right) u(e^{it}) \, dt \quad (|z| < 1).$$

4.   Let $\gamma(t) = e^{it}$ $(t \in [-\frac{1}{2}\pi, \frac{1}{2}\pi])$ and let $f(z) = \int_\gamma (w - z)^{-1} \, dw$ $(z \notin \gamma^*)$. Compute $f^{(n)}(0)$ for $n = 0, 1, 2, \ldots$, and deduce that f has a Taylor series expansion

$$2\pi i f(z) = \frac{1}{2} + \frac{1}{\pi} \sum_{n=0}^{\infty} (-1)^n z^{2n+1}/(2n + 1),$$

valid for $|z| < 1$.

5.   (i) Is it possible to find a function f holomorphic in $D(0; 1)$ such that

$$f(1/n) = f(-1/(n + 1)) = 1/(n + 1) \quad (n = 1, 2, 3, \ldots)?$$

   (ii) Is it possible to find a function f holomorphic in $D(0; 1)$ such that

$$f(1/n) = f(-1/n) = 1/n^3 \quad (n = 1, 2, 3, \ldots)?$$

   (iii) Find two different functions f_1, f_2, each holomorphic in $\mathbb{C}$, such that $f_1(n/2) = f_2(n/2) = n^2$ for $n = 1, 2, 3, \ldots$.

6.   Let $\langle \alpha_n \rangle$ be a sequence of complex numbers such that $\Sigma |\alpha_n|$ converges and $\Sigma_{n=1}^{\infty} \alpha_n k^{-n} = 0$ for all $k = 1, 2, 3, \ldots$. Prove that $\alpha_n = 0$ for all n.

7.   Assume there exists a function f which is holomorphic in the region $\{z : \text{Re } z > 0\}$ and which satisfies

$$f(z + 1) = zf(z) \quad (\text{Re } z > 0).$$

Show that f_1, defined by $f_1(z) = f(z + 1)/z$, is holomorphic in $\{z : \text{Re } z > -1, z \neq 0\}$ and satisfies $f_1(z + 1) = zf_1(z)$. Show further that there is a function g such that
   (i) g is holomorphic in $G := \mathbb{C} \backslash \{0, -1, -2, \ldots\}$,
   (ii) $g(z) = f(z)$ for $\text{Re } z > 0$, and
   (iii) $g(z + 1) = zg(z)$ for $z \in G$.

8.   Let f be holomorphic in $\mathbb{C}$. Prove that either of the following conditions forces f to be a constant function:
   (i) $\text{Re } f \leq M$ (where M is a finite constant),
   (ii) $f(z)$ is real when $|z| = 1$.
(Hint: consider suitable exponentials.)

9.   Use Supplementary Exercise 3.5 to give alternative proofs of
   (i) Liouville's theorem, 5.2,
   (ii) the Local maximum-modulus theorem, 5.19.

10.   Suppose that $f \in H(D(0; 1))$ is such that $|f(z)| < 1$ for all $z \in D(0; 1)$

and $f(0) = 0$. Show that there exists $g \in H(D(0; 1))$ such that $f(z) = zg(z)$. Apply the Maximum-modulus theorem to g in $D(0; r)$ $(r < 1)$ to prove that

$$|f(z)| \leq |z| \quad (|z| < 1) \quad \text{and} \quad |f'(0)| \leq 1,$$

with strict inequalities unless $f(z) = cz$ where c is a constant of modulus 1. (This is *Schwarz's lemma*.)

11. Let $U = \{z : \text{Im } z > 0\}$. Suppose $F : U \to U$ is holomorphic and that $a \in U$. Prove that, for all $z \in U$,

$$\left| \frac{F(z) - F(a)}{F(z) - \overline{F(a)}} \right| \leq \left| \frac{z - a}{z - \bar{a}} \right| \quad \text{and} \quad |F'(a)| \leq \frac{\text{Im } F(a)}{\text{Im } a}.$$

(Hint: consider the composite function $f = \varphi \circ F \circ \varphi^{-1}$, where $\varphi(z) = (z - a)/(z - \bar{a})$, and apply the results of the previous exercise.)

Chapter 6

1. Let f and g be holomorphic in $D(a; r)$ for some $r > 0$ and assume that $f(a) = g(a) = 0$. Use 6.9 to prove that

$$\lim_{z \to a} \frac{f(z)}{g(z)} = \lim_{z \to a} \frac{f'(z)}{g'(z)},$$

if the right-hand side exists. Use this complex form of L'Hôpital's rule to evaluate

(i) $\lim_{z \to i} \dfrac{1 + e^{\pi z}}{1 + z^2}$, (ii) $\lim_{z \to 0} \cot z - z^{-1}$,

(iii) $\lim_{z \to 1} \dfrac{(1 - z)}{1 - z e^{\lambda(1 - z)}}$ (λ constant).

2. By considering the Laurent expansions of $(z + z^{-1})^m$ and $(z - z^{-1})^m$ and integrating suitable functions round the unit circle, prove that, for $m, k \in \mathbb{Z}$ and $0 \leq m < |k|$,

$$\int_0^{2\pi} (\cos \theta)^m e^{ik\theta} \, d\theta = \int_0^{2\pi} (\sin \theta)^m e^{ik\theta} \, d\theta = 0.$$

Use a similar technique to prove that, for $p = 1, 2, 3, \ldots$,

$$\int_0^{2\pi} (\cos \theta)^{2p} \, d\theta = 2\pi \frac{(2p)!}{2^{2p} p!^2}.$$

3. Let f be holomorphic in $\mathbb{C}$.

(i) Prove that f has a removable singularity at ∞ if and only if it is a constant.

(ii) Prove that f has a pole of order m at infinity if and only if it is a polynomial of degree m.

Characterize (a) those rational functions which have a removable singularity at ∞ and (b) those rational functions which have a pole at ∞.

4. Assume that f has a pole of order m at a and that g has a pole of order n at a. What kind of singularity at a is it possible for (i) $f + g$, (ii) $f \circ g$ to have? Give examples to show that all the possibilities you list can occur.

5. Determine the type of singularity that each of the following functions has at (a) 0, (b) ∞:

(i) $\left(z + \dfrac{1}{z}\right)^{-1}$,   (ii) $z^2 e^{1/z}$,   (iii) $\cot z - \dfrac{1}{z}$,   (iv) $(z - \sin z)^{-1}$,

(v) $\operatorname{cosec}(\sin z)$.

6. Let G be a bounded region in $\mathbb{C}$ and let S be a closed subset of $\mathbb{C}$ contained in G. Is it possible to construct

(i) a function which has an infinite number of poles in S and is otherwise holomorphic in G;

(ii) a function f which is holomorphic in G and is such that $1/f$ has an infinite number of poles in S?

7. [This exercise assumes familarity with uniform convergence.] Define
$$f(z) = \frac{1}{z} - 2z \sum_{n=1}^{\infty} \frac{(-1)^n}{n^2 \pi^2 - z^2}.$$

(i) Prove that f is holomorphic in $G := \mathbb{C} \backslash \{n\pi : n \in \mathbb{Z}\}$, by showing that the series converges uniformly on any disc $D(a; r) \subseteq G$ and applying the result in Exercise 5.15.

(ii) Prove that f has a simple pole at each point $k\pi$ ($k \in \mathbb{Z}$). (Hint: split $f(z)$ into two parts, a finite sum having a pole at $k\pi$ and an infinite sum holomorphic at $k\pi$.)

(iii) Deduce that $f(z) = \operatorname{cosec} z$ for all $z \in G$.

8. Let $a_1, \ldots, a_m$ be distinct points of $\mathbb{C}$. Prove that the multifunction $[[(z - a_1) \ldots (z - a_m)]^{\frac{1}{2}}]$ has branch points at each point a_k and has a branch point at ∞ if k is odd but not if k is even.

9. For each of the following multifunctions, locate the branch points, suggest how the plane should be cut, and specify a holomorphic branch:
(i) $[[z^2(1 - z)]^{\frac{1}{2}}]$, (ii) $[[z + 1/z]^{\frac{1}{2}}]$, (iii) $[[(z - 1)(z - w)(z - w^2)]^{\frac{1}{2}}]$, where $w = e^{2\pi i/3}$, (iv) $[\log(\sqrt{z})]$ ($:= \{\log |z|^{\frac{1}{2}} + \frac{1}{2}i\theta : \theta \in [\arg z]\}$).

10. In the plane cut along $(-\infty, 0]$, determine the square root $f(z)$ of z by $|z|^{\frac{1}{2}} e^{i\theta/2}$ ($0 \neq z = |z| e^{i\theta}$, $-\pi < \theta \leqslant \pi$). By using the identity
$$f(z + h) - f(z) = \frac{h}{f(z + h) + f(z)},$$

prove that $f'(z)$ exists and equals $f(z)/2z$ in $\mathbb{C}\backslash(-\infty, 0]$. By considering

$$\frac{d}{dz}\left(\frac{1}{f(z)}\sum_{n=0}^{\infty}\frac{\frac{1}{2}(\frac{1}{2}-1)\ldots(\frac{1}{2}-n+1)}{n!}(z-1)^n\right)$$

for $|z-1|<1$, or by assuming an expansion for $x^{\frac{1}{2}}$ for suitable real x, obtain the Taylor expansion of f in $D(1; 1)$.

Chapter 7

1. Calculate the residues at the poles in $\mathbb{C}$ of the following functions:

(i) $\dfrac{z^2}{(z^2+1)^2}$,   (ii) $\dfrac{1}{z^2-z^6}$,   (iii) $\pi\tan\pi z$,   (iv) $\dfrac{ze^{iz}}{z^6+1}$.

2. Let $f, g \in H(D(a; r))$ and assume that f has a zero of order m and g a zero of order $m+1$ at a. Prove that

$$\text{res}\left\{\frac{f(z)}{g(z)}; a\right\} = (m+1)\frac{f^{(m)}(a)}{g^{(m+1)}(a)}.$$

3. The function f is holomorphic in $\check{\mathbb{C}}$ except for simple poles at ± 2, the residue at 2 being $\frac{1}{2}$. Determine f, given that it has a zero of order 2 at 0.

4. Let f be holomorphic inside and on $\gamma(a; r)$ and assume that $f(z)\neq 0$ for $z \in \gamma(a; r)^*$. Find, in terms of the zeros of f, the values of the integral

$$\frac{1}{2\pi i}\int_\gamma \frac{f'(z)}{f(z)} z^m \, dz,$$

for $m = 1, 2, 3, \ldots$. (Theorem 7.6 treats the case $m = 0$.)

5. Determine the number of zeros of each of the following functions in $D(0; 1)$:
   (i) $z^5 - 3z + 1$,
   (ii) $z^7 + 2z^5 + 2z^2 + 6$,
   (iii) $\cos\pi z - 100z^n$.

6. Show that, for each $\lambda > 1$, the equation $z + e^{-z} = \lambda$ has precisely one zero in the open right half-plane, and that this zero is real.

7. Suppose that f is holomorphic inside and on $\gamma(0; 1)$, and has Taylor series $\sum_{n=0}^{\infty}c_n z^n$. Given that f has m zeros inside $\gamma(0; 1)$, prove that

$$\min\{|f(z)| : |z| = 1\} \leqslant |c_0| + |c_1| + \ldots + |c_m|.$$

Chapter 8

1. Evaluate by contour integration

(i) $\displaystyle\int_0^\infty \frac{x^2}{(x^2+a^2)^3}\,dx \quad (a>0),$

(ii) $\displaystyle\int_0^\infty \frac{\cos ax}{(x^2+b^2)^2}\,dx \quad (a,\,b>0),$

(iii) $\displaystyle\int_0^\infty \frac{\log x}{(x-a)^2+b^2}\,dx \quad (a,\,b>0).$

2. By integrating round a suitable sector prove that for $s \in \mathbb{R}$, $\lambda>0$ and $n = 1, 2, 3, \ldots,$

$$\int_0^\infty x^{n-1}e^{-\lambda x - isx}\,dx = \frac{(n-1)!}{(\lambda+is)^n}.$$

3. State, with reasons, what function f and contour γ you would choose in order to evaluate, by applying Cauchy's theorem or Cauchy's residue theorem to $\int_\gamma f(z)\,dz$, the following integrals:

(i) $\displaystyle\int_{-\infty}^\infty \frac{x \sin mx}{1+x^2}\,dx \quad (m \in \mathbb{R}),$ (ii) $\displaystyle\int_0^\infty \frac{1-\cos x}{x^2(1+x^4)}\,dx,$

(iii) $\displaystyle\int_0^\infty \frac{x}{\sinh x}\,dx,$ (iv) $\displaystyle\int_{-\infty}^\infty \frac{x^2}{1+x^{12}}\,dx,$

(v) $\displaystyle\int_0^\infty \frac{1}{(1+x^2)^n}\,dx,$ (vi) $\displaystyle\int_{-\infty}^\infty \frac{e^{ax}\cos x}{1+e^x}\,dx \quad (-1<a<1),$

(vii) $\displaystyle\int_{-\infty}^\infty \frac{e^{-ix}}{1+x^4}\,dx,$ (viii) $\displaystyle\int_0^{2\pi} \frac{1}{a+b\sin x}\,dx \quad (a,\,b \in \mathbb{R},\,a>|b|).$

Specify the location of any singularities of the functions and comment on how you would obtain any limit integrals.

4. Repeat the previous exercise for the following integrals, specifying an appropriate holomorphic branch of any multifunction used.

(i) $\displaystyle\int_0^\infty \frac{x^a}{(1+x)^2}\,dx \quad (-1<a<1),$ (ii) $\displaystyle\int_0^\infty \frac{\log(1+x^2)}{x^2}\,dx,$

(iii) $\displaystyle\int_0^\infty \frac{\log x}{x^2+x-2}\,dx,$ (iv) $\displaystyle\int_0^\infty \frac{\sin\sqrt{x}}{x(1+x^2)}\,dx.$

5. Let γ be the rectangle with vertices at $0, R, R+i\pi, i\pi$. By integrating round γ, suitably modified by indentations, prove that

$$\int_0^\infty \frac{\sin ax}{e^{2x}-1}\,dx = -\frac{1}{2a}+\frac{\pi}{4}\coth\frac{\pi a}{2}.$$

6. Prove that

(i) $\displaystyle\sum_{n=-\infty}^{\infty} (n-a)^{-2} = \pi^2 \operatorname{cosec}^2 \pi a \quad (a \text{ real}, a \notin \mathbb{Z}).$

(ii) $\displaystyle\sum_{n=-\infty}^{\infty} (-1)^n (n^2+1)^{-1} = \pi \operatorname{cosech} \pi.$

7. Derive the following expansions:

(i) $\displaystyle \pi \cot \pi z = \frac{1}{z} + \sum_{n=1}^{\infty} \frac{2z}{z^2 - n^2},$

(ii) $\displaystyle \frac{1}{e^z - 1} = \frac{1}{z} - \frac{1}{2} + \sum_{n=1}^{\infty} \frac{2z}{z^2 + 4n^2\pi^2} \quad (z \neq 2m\pi i \ (m \in \mathbb{Z})).$

Chapter 9

1. Prove by induction that

$$\mathscr{L}[(1-e^{-\alpha t})^n] = \frac{n!\alpha^n}{p(p+\alpha)\dots(p+n\alpha)} \quad (n = 0, 1, 2, \dots).$$

Hence compute $\mathscr{L}[\sin^n t]$.

2. Find the inverse Laplace transform of each of (i) $2p^2/(p^4+1)$, (ii) $1/((p^2+4)(p^2+1)^2)$, (iii) $p^{-\frac{3}{2}}e^{-1/p}$ (where $p^{-\frac{3}{2}}$ denotes a suitable holomorphic branch of $[p^{-\frac{3}{2}}]$).

3. Let $f(x) = \left(1 - \dfrac{|x|}{\lambda}\right) \chi_{[-\lambda, \lambda]}(x)$. Compute the Fourier transform of f and hence find the value of

$$\int_{-\infty}^{\infty} \frac{\sin^2 x}{x^2} \, dx.$$

4. Use the Laplace transform to solve

$$\frac{d^2y}{dt^2} - 5\frac{dy}{dt} + 4y = \begin{cases} 1 & \text{if } 0 < t \leq b, \\ -1 & \text{if } b < t \leq 2b, \\ 0 & \text{if } 2b < t, \end{cases}$$

with $y(0) = 0$, $y'(0) = 1$.

5. Let $f(t) = t^{-a}$, where $0 < a < 1$. For $\operatorname{Re} p > 0$, compute the Laplace transform $\bar{f}(p)$ in terms of the constant C_a, where

$$C_a = \int_0^{\infty} u^{-a} e^{-u} \, du.$$

Use 5.18 and 9.5 to obtain $\bar{f}(p)$ for $\operatorname{Re} p > 0$. Apply the Inversion theorem to prove that

$$C_a C_{1-a} = \pi \operatorname{cosec} \pi a.$$

6. Suppose that f satisfies $f'(t) = f(kt)$ $(t > 0)$, where $0 < k < 1$, and $f(0) = 1$. Prove that

$$f(t) = \sum_{n=0}^{\infty} \frac{k^{n(n-1)/2}}{n!} t^n.$$

7. Use the Fourier transform to solve

$$\frac{d^2 f}{dx^2} + 2 \frac{df}{dx} + f(x) = g(x) \qquad (x \in \mathbb{R}),$$

expressing the solution as a convolution.

8. The function $u(x, t)$ is continuous in $\{(x, t) : x \geqslant 0, t \geqslant 0\}$ and satisfies

(i) $u_{tt} = c^2 u_{xx}$ $\qquad (x > 0, t > 0)$,

(ii) $u(x, 0) = u_t(x, 0) = 0$ $\qquad (x > 0)$,

(iii) $\dfrac{d^2}{dt^2} u(0, t) + \mu^2 u(0, t) = \dfrac{2c^2}{b} u_x(0, t)$ $\qquad (\mu > c/b)$,

(iv) $u(0, 0) = 0$, $\qquad \left[\dfrac{d}{dt} u(0, t)\right]_{t=0} = U$,

where c, μ, b, and U are constants. Making such technical assumptions as you need, obtain $u(x, t)$, and verify that the solution you have found does satisfy the given conditions and any additional conditions imposed.

9. Given $f \in \mathscr{S}([0, \infty))$, extend f to $F \in \mathscr{S}(\mathbb{R})$ by defining

$$F(x) = \begin{cases} f(x) & (x \geqslant 0), \\ -f(-x) & (x < 0). \end{cases}$$

By applying the Inversion theorem for the Fourier transform to F, show that, for suitably smooth f, $f(t)$ can be expressed in terms of its sine transform:

$$\bar{f}(s) := \int_0^\infty f(t) \sin ts \, dt.$$

Use the sine transform to give an alternative solution to Example 9.23.

Chapter 10

1. Find a Möbius transformation mapping G_1 onto G_2 when
   (i) $G_1 = \{z : \operatorname{Im} z < 1/2\}$, $G_2 = D(0; 1)$,

(ii) $G_1 = \{z : -\pi/2 < \arg z < \pi/2\}$, $G_2 = \{w : |w| < 1,\ \text{Im } w < 0\}$,

(iii) $G_1 = \{z : 0 < \text{Im } z < 1\}$, $G_2 = \{w : |w - 1/2| > 1/2,\ \text{Re } w < 1\}$,

2. (i) Find the image of $U := \mathbb{C}\backslash\{z : |z| = 1,\ \text{Im } z \geqslant 0\}$ under $z \mapsto (z - i)/(z + i)$.

(ii) Find the image of $\mathbb{C}\backslash[-1, 1]$ under $z \mapsto (z - 1)/(z + 1)$.

(iii) Hence find a conformal mapping of U onto a half-plane.

3. Show that $z \mapsto \zeta := i\dfrac{1 + z}{1 - z}$ maps the unit circle onto the real axis and

deduce that the same is true of the map $z \mapsto \zeta^3$.

Determine and sketch the regions in the z-plane that are carried onto the right half-plane $\{w : \text{Re } w > 0\}$ by (i) $z \mapsto \zeta$, (ii) $z \mapsto \zeta^3$.

4. Let $f : z \mapsto w = f(z) = \dfrac{az + b}{cz + d}$ be a Möbius transformation. A point $\alpha \in \mathbb{C}$ is said to be a *fixed point* of f if $f(\alpha) = \alpha$.

(i) Prove that f has either one or two fixed points.

(ii) Suppose f has distinct fixed points, α and β. Prove that

$$\frac{w - \alpha}{w - \beta} = k\frac{z - \alpha}{z - \beta}, \quad \text{where } k = \frac{a - c\alpha}{a - c\beta}.$$

What is the image under f of (a) the circline $|z - \alpha|/|z - \beta| = \lambda$, (b) the circular arc $\arg(z - \alpha) - \arg(z - \beta) = \mu \pmod{2\pi}$?

(iii) Suppose f has a single fixed point, α. Prove that

$$\frac{1}{w - \alpha} = \frac{1}{z - \alpha} + K \quad \text{where } K = \frac{c}{a - c\alpha}.$$

What is the image under f of the circle $|z - \alpha| = r$?

5. Suppose that f is a Möbius transformation mapping the open upper half-plane onto itself. How are $f(i)$ and $f(-i)$ related? Determine the general form f may take.

6. Find a map built from exponentials to map G_1 conformally onto G_2 when

(i) $G_1 = \{z : \text{Im } z > 0\}$, $G_2 = \{w : |w| > 1\}$,

(ii) $G_1 = \{z : -\pi/6 < \text{Im } z < \pi/6\}$, $G_2 = \{w : \text{Re } w > 0\}$,

(iii) $G_1 = \{z : |\text{Re } z| < \pi/2,\ \text{Im } z > 0\}$, $G_2 = \{w : \text{Im } w > 0\}$.

In each case specify also a conformal map from G_2 onto G_1.

7. Find the image of the strip $\{z : 0 < \text{Re } z < \pi/2\}$ under the map $z \mapsto \operatorname{cosec}^2\left(\dfrac{\pi}{4} + \dfrac{z}{2}\right)$.

8. Let G be an open set and let γ be a path with $\gamma^* \subseteq G$. Suppose that $f : G \to \mathbb{C}$ is a function such that the partial derivatives f_x, f_y exist and are

continuous in G and let $\Gamma = f \circ \gamma$ be the image of γ under f. Show that

$$\Gamma'(t) = \frac{1}{2}(f_x - if_y)\gamma'(t) + \frac{1}{2}(f_x + if_y)\overline{\gamma'(t)},$$

where the partial derivatives are evaluated at $\gamma(t)$. By considering $\arg[\Gamma'(t)/f'(t)]$, for suitable choices of γ, show that, if f preserves the magnitude and sense of angles between paths in G, then the real and imaginary parts of f satisfy the Cauchy–Riemann equations. (Hence, by 10.30, $f \in H(G)$.)

Bibliography

1. Apostol, T. M., *Mathematical analysis* (1st edn). Addison-Wesley, Reading, Mass. (1957).
2. Apostol, T. M., *Mathematical analysis* (2nd edn). Addison-Wesley, Reading, Mass. (1974).
3. Beardon, A. F., *Complex analysis.* Wiley, New York (1979).
4. Binmore, K. G., *Mathematical analysis* (2nd edn). Cambridge University Press, Cambridge (1982).
5. Hille, E., *Analytic function theory*, Vol. I. Ginn and Company, Boston (1959).
6. Kosniowski, C., *A first course in algebraic topology.* Cambridge University Press, Cambridge (1980).
7. Rudin, W., *Real and complex analysis* (2nd edn). McGraw Hill, New York (1974).
8. Weir, A. J., *Lebesgue integration and measure.* Cambridge University Press, Cambridge (1973).

Further supplementary and collateral reading

1. Ahlfors, L. V., *Complex analysis* (2nd edn). McGraw-Hill, New York (1966).
2. Ash, R. B., *Complex variables.* Academic Press, New York, London (1971).
3. Beardon, A. F., *A primer on Riemann surfaces.* Cambridge University Press, Cambridge (1984).
4. Burckel, R. B., *An introduction to classical complex analysis*, Vol. I. Birkhäuser Verlag, Basel (1979).
5. Hofmann, K., *Banach spaces of analytic functions.* Prentice-Hall, Englewood Cliffs, N.J. (1962).
6. Knopp, K. *Theory of functions* (translated by F. Bagemihl), Vols I, II. Dover Publications (1975).
7. Levinson, N. and Redheffer, R. M., *Complex variables.* Holden-Day, San Francisco (1970).
8. Silverman, R. A., *Complex analysis with applications.* Prentice-Hall, Englewood Cliffs, N.J. (1974).

9. Stewart, I. and Tall, D., *Complex analysis*. Cambridge University Press, Cambridge (1983).
10. Titchmarsh, E. C., *The theory of functions* (2nd edn). Oxford University Press, Oxford (1939).
11. Watson, E. J. *Laplace transforms and applications*. Van Nostrand Reinhold, New York (1981).

Notation Index

Subject Index